Pesticides - Use and Misuse and Their Impact in the Environment

Edited by Marcelo Larramendy and Sonia Soloneski

Published in London, United Kingdom

IntechOpen

Supporting open minds since 2005

Pesticides - Use and Misuse and Their Impact in the Environment
http://dx.doi.org/10.5772/intechopen.78909
Edited by Marcelo Larramendy and Sonia Soloneski

Contributors
Bunini Manyilizu, Tamara García-Barrera, José Luis Gómez-Ariza, Gema Rodríguez-Moro, Ana Arias-Borrego, Sara Ramírez-Acosta, Francisco Navarro-Roldán, Nieves Abril, Rut Fernández-Torres, Miguel-Angel Bello-Lopez, Simón Navarro, Gabriel Pérez-Lucas, Nuria Vela, Abderrazak El Aatik, Liliana Cristina Soare, Cristina Maria Ponepal, Alina Păunescu, Mona AL-Ahmadi, Pouokam Guy Bertrand

First published in London, United Kingdom, 2019 by IntechOpen
IntechOpen is the global imprint of INTECHOPEN LIMITED, registered in England and Wales, registration number: 11086078, The Shard, 25th floor, 32 London Bridge Street
London, SE19SG - United Kingdom
Printed in Croatia

British Library Cataloguing-in-Publication Data
A catalogue record for this book is available from the British Library

Additional hard and PDF copies can be obtained from orders@intechopen.com

Pesticides - Use and Misuse and Their Impact in the Environment
Edited by Marcelo Larramendy and Sonia Soloneski
p. cm.
Print ISBN 978-1-83962-647-0
Online ISBN 978-1-83880-047-5
eBook (PDF) ISBN 978-1-83880-048-2

For EU product safety concerns:
IN TECH d.o.o., Prolaz Marije Krucifikse Kozulić 3, 51000 Rijeka, Croatia, info@intechopen.com or visit our website at intechopen.com.

Meet the editors

Marcelo L. Larramendy, PhD, serves as a professor of molecular cell biology at the School of Natural Sciences and Museum (National University of La Plata, Argentina). He was appointed senior researcher of the National Scientific and Technological Research Council of Argentina and was a former member of the Executive Committee of the Latin American Association of Environmental Mutagenesis, Teratogenesis and Carcinogenesis. He is the author of more than 450 contributions, including scientific publications, research communications, and conference papers worldwide. He is the recipient of several national and international awards. Prof. Larramendy is a regular lecturer at international A. Hollaender Courses organized by the IAEMS and is a former guest scientist at NIH (USA) and the University of Helsinki (Finland). He is an expert in genetic toxicology and has been, a referee for more than 20 international scientific journals. He was a member of the International Panel of Experts at the International Agency for Research on Cancer (WHO, Lyon, France) in 2015 for the evaluation of DDT, 2,4-D, and lindane. Presently, Prof. Dr. Larramendy is the head of the Laboratory of Molecular Cytogenetics and Genotoxicology at the UNLP.

Sonia Soloneski has a PhD in Natural Sciences and is an assistant professor of molecular cell biology at the Faculty of Natural Sciences and Museum of La Plata, National University of La Plata, Argentina. She became a member of the National Scientific and Technological Research Council of Argentina in the field of genetic toxicology. Presently, she is a member of the Latin American Association of Environmental Mutagenesis, Teratogenesis and Carcinogenesis, the Argentinean Society of Toxicology, and the Argentinean Society of Genetics. She has authored more than 350 scientific publications, including scientific publications in research papers, reviewed journals, and conference papers worldwide. She is a referent for issues in the field of genetic toxicology, mutagenesis, and ecotoxicology.

Contents

Preface

Throughout human history, pest control has been continually associated with man-made activities and it has undoubtedly played a role in protecting crops. For two thousand years, several kinds of pesticides have been used, not only by farmers but also by others, to control pests, weeds, and diseases. It is well known that the ancient Egyptians employed compounds containing alkaloids, including, among others, hemlock, aconite, and opium, to control pests, which were adopted as great favorites of the Greeks and Romans for use as poisons employed in medicine, rituals, and even war. During the industrial revolution, significant advances in the manufacture of synthetic chemicals were developed. Some chemicals employed during World Wars I and II accelerated the development of the modern chemical industry. Synthetic chemicals and technologies originally employed for warfare were then modified and improved for several civilian uses. In 1939, the Swiss chemist Paul Müller designed the insecticide dichlorodiphenyltrichloroethane, commonly known as DDT, which was manufactured to eliminate unwanted insects. However, it is also known that the insecticide was definitively banned in 1972 due to its severe toxicity. German researchers in 1943 developed organophosphate compounds, such as parathion, with insecticide properties, which are still in use today despite their high and persistent toxicity. Presently, international and national government agencies and administrations regulate the manufacture and use of pesticides, which are developed to persist in the environment for shorter periods and be less toxic for nontarget organisms to reduce environmental risk as a consequence of their use as pest control.

Widespread pesticide use continues to be indispensable for maintaining sustainable agriculture, the control of pest-borne diseases, and the health of human populations and ecological systems, among other uses. Consequently, pesticides are repeatedly introduced throughout the environment and they make a great contribution to the pollution of the environment since the global distribution is the manner they are employed.

Although a vast literature is available on pesticides, this book contains relevant information on diverse pesticides encountered in both anthropogenic and natural environments and provides valuable information on the toxicity of several agrochemicals that can negatively influence the health of humans and ecosystems.

The book begins with a chapter presenting an approach of several biomarkers employed for determining pesticide pollution. Different aspects have been deeply analyzed throughout the chapter, including the influence of diverse pesticides on the spore germination process and the differentiation of their gametophyte on ferns; the impact of some pesticides on fish breathing at physiological, histopathological, and tissue levels; as well as the toxicological pattern at hematological, biochemical, and structural parameters in amphibians. The second chapter is focused on the *omics* analytical methodologies as efficient current tools for evaluating the final biological response exerted by several environmental pollutants, e.g. chemical mixtures. The third chapter provides information on how several processes, namely leaching, diffusion, volatilization, erosion and run-off, assimilation by microorganisms, as well as plant uptake, can displace different types of pesticides throughout the soil profile to increase groundwater pollution. The fourth chapter is an interesting review showing how massive pesticides—synthetic pesticides, biopesticides, and nanopesticides—are employed continually to protect crops, highlighting the harmful side effects inflicted on human populations. The fifth chapter provides information on the uses and misuses of agricultural pesticides in Africa. Lastly, the final chapter presents a review of the historical events related to pesticide employment in Africa. The chapter focuses on human activities that require the use of chemical agents for pest control to protect crops and animals, including humans, in African countries.

The editors of *Pesticides - Use and Misuse and Their Impact in the Environment* are enormously grateful to all the authors for contributing to this book. They have made an extensive effort to arrange the information included in every chapter. The contributions made by specialists in this field of research are gratefully acknowledged. We hope that the information presented in this book will continue to meet the expectations and needs of all those interested in the different aspects of pesticides.

The publication of this book is of great importance to those scientists, pharmacologists, physicians, and veterinarians, as well as engineers, teachers, graduate students, and administrators of environmental programs, who make and employ different investigations to understand both basic and applied aspects of the use and misuse of pesticides.

Sonia Soloneski PhD and Marcelo L. Larramendy PhD
School of Natural Sciences and Museum,
National University of La Plata,
La Plata, Argentina

Chapter 1

The Morphophysiological, Histological, and Biochemical Response of Some Nontarget Organisms to the Stress Induced by the Pesticides in the Environment

Liliana Cristina Soare, Alina Păunescu and Ponepal Cristina Maria

Abstract

Ferns, amphibians, and fish are groups of nontarget organisms affected by many types of pesticides that end up in the environment. This chapter aims to approach the following themes: the influence of different pesticides on the spore germination process and on the differentiation of their gametophyte; aspects regarding the impact of some pesticides on breathing in fish (physiology and histopathology at the branchial level), as well as a series of effects at the hematological and biochemical levels; and changes of some hematological, biochemical, and structural parameters in amphibians. Species that are not directly targeted by the action of the pesticide in the environment, ferns can be used in their gametophyte stage, young or mature sporophyte in different biotests to evaluate the risk associated with these substances. The biochemical, hemathological, and histopathological changes recorded in both fish and amphibians can be considered biomarkers of pesticide pollution.

Keywords: fern, fish, amphibians, pesticides

1. Introduction

Contamination of the natural environment, due to natural factors and human action, has been subject of numerous studies. In many ecosystems, human activity has led to changes of natural biogeochemical cycles, resulting in the accumulation of some substances.

Chemical contaminants are found everywhere in nature, and ecologists are the ones who assess their impact on natural communities of organisms. Among these, pesticides are the most common type of aquatic ecosystem contaminant.

Considering the physical and chemical properties (vapor expansion, volatility, evaporation capacity from water—codistillation) that transform pesticides, their bioaccumulative stability and capacity as well as the publication of ecotoxicological data on the action of different pesticides have become indispensable to continually monitor changes in the environment.

Different research on morphology, physiology, and biochemistry of gametophyte and sporophyte of fern species has shown that pesticides affect these processes, depending on species sensitivity, concentration, and exposure time.

Fish assimilates pesticides through gills or contaminated food. Gills are the main channel of pesticide penetration, which is why any disease at this level will have a great influence on the adaptive changes in the fish body. The effects of pesticides on fish are numerous and varied: they cause mortality both directly and indirectly, by starvation (destroying the organisms they feed on), affects hatching, growth rate, can lead to malformations, affects reproduction rate, modifies enzyme activity, and cause histopathological changes in organs, genetic effects, etc. Although under experimental conditions fish can survive pesticides in different concentrations, under natural conditions they are more vulnerable to disease, predators, no longer competitive, and no longer dealing with stress caused by changing seasonal temperature, reproduction season or temporary starvation.

Amphibians are organisms that populate aquatic ecosystems, being involved in aquatic trophic chains both by eating food and being food for predators. Over the last 20 years, scientists have reported the global decline of amphibian populations. Therefore, in 1996, the International Union for Conservation of Nature registered 156 species of amphibians in the Red List, and recent data show that 1856 species (about 32.5% of all amphibian species) are currently registered.

2. Influence of pesticides on spore germination and gametophyte differentiation in ferns

2.1 Gametophyte of ferns

Ferns and lycophytes (pteridophytes) represent 4% of Terra's vascular plants [1], numbering about 11,000 species in 40 families and 300 genres [2]. Different species of pteridophytes are used as medicinal, food, horticultural, and agricultural plants [1] and in the last decade as organisms in acute [3] and chronic phytotoxicity tests [4].

The life cycle of pteridophytes is characterized by an alternation of generations between a well-developed sporophyte and a reduced gametophyte, independent of the sporophyte [1]. The gametophytic generation of ferns begins with the formation of spores (meiospores), from which, a gametophyte or multilcellular prothallium is differentiated by germination. The gametophyte of leptosporangiate ferns is above ground, photosynthetic, short lived, and heart shaped [1]. From the spore to the mature gametophyte stage with gamentagia, the differentiation of the gametophyte involves the passage through the prothallium filament and spatula-shaped prothallium lamella [5].

The spores germinate on the ground, the first division resulting in a smaller base cell (the initial cell of the rhizoid) and a larger apical cell (the initial cell of the prothallium). The initial cell of the rhizoid elongates and forms a hyaline rhizoid having no chloroplasts. The initial cell of the prothallium will form a multicellular prothallium filament, consisting of a variable number of cells with numerous chloroplasts. One or more rhizoids are formed on the cells at the base of the prothallium filament. Lamella formation (prothallium plate) is initiated by the longitudinal division of the end cell of the prothallium filament. One of the daughter cells divides through a slanted wall resulting in an apical (initial) wedge-shaped cell. The fore side of the prothallium then extends to a width of 3–4 cells and becomes spatula shaped. When the apex of the prothallium is heart shaped, the initial cell is replaced by a flat multicellular meristem. The activity of the lateral zones of this multicellular meristem helps the formation of the prothallium wings. The cells behind the apical meristem divide through a plan parallel to the underlayer, leading to the formation of a multi-layered medial crest. On the underside of the prothallium, there are antheridia, archegonia, and rhizoids. Prothallium differentiates different types of prothallium trichomes in some species.

The processes of spore germination and differentiation of gametophyte in ferns are influenced by a series of endogenous and exogenous factors, with environmental pesticides in the latter category.

2.2 Changes in the fern gametophyte caused by pesticides

Over the time, many researchers have highlighted the advantages of using spores and gametophytes of ferns in their experimental research [6] and ecotoxicology in recent years [3, 7]. The most advantages include: (1) the formation of a large number of spores/individual, (2) spores can be preserved in the laboratory and can thus be available throughout a year (3) are small and light, requiring reduced storage space, (4) spore germination and gametophyte differentiation may be obtained and monitored *in vitro* in small recipients, in simple laboratory conditions, and (5) conclusions are relevant for higher plants.

The spore germination process can be completely inhibited or delayed by pesticides in the environment. The glyphosate herbicide in concentrations of 0.48–19.20 mg/l significantly inhibited the macrospores germination process of *Regnellidium diphyllum* aquatic fern. From a total of 1050 megaspores used in the experiment, only 744 germinated [8].

A fungicide containing 50% metal copper applied in a concentration of 0.1–0.3 g/100 ml Knop solution significantly inhibited the spore germination process in *Athyrium filix-femina* and *Polypodium vulgare* ferns. The highest concentration of fungicide in the first species inhibited germination completely [9]. After 50 days of exposure to the fungicide, gametophyte differentiation was delayed compared to the control variant, which was in young cordate prothallium stage. Except for the lowest concentration of fungicide applied to *P. vulgare* in which the gametophyte was in the stage of prothallium lamella, at all the others, the stage of differentiation was prothallium filament. In addition, the filament cells had necroses, and the rhizoids were unelongated. Also *P. vulgare* species had more sinuous and more intensely colored cell walls at the level of rhizoids, [9] compared to the control variant.

A bifenthrin-based insecticide applied in a concentration of 0.01–0.04 ml/100 ml of Knop solution to the two above-mentioned species resulted in a decrease in spore germination rates, and at the highest concentration in both species, the gametophyte was in the form of a three-dimensional cell mass, whereas the gametophyte in the control variant was in cordate prothallium stage [9].

Asplenium scolopendrium spores which were grown *in vitro* on 20% copper metal fungicide containing culture medium showed a decrease in the germination percentage by up to 37.66% compared to the control variant. The species proved more sensitive than *Athyrium filix-femina* when the highest fungicide concentration was applied [10]. Gametophyte differentiation of both species was affected, so after 3 weeks of exposure, the gametophyte was in the stage of prothallium lamella formation for *A. scolopendrium* control variant, respectively, in the stage of young cordate prothallium for *A. filix-femina* control variant. The gametophyte was in the form of germinated spores when the highest concentration of fungicide (0.7%) was applied. Necrosis of prothallic chlorocytes, inhibition of rhizoid elongation, and even the absence of rhizoids have been observed in variants exposed to fungicide.

Acetamiprid-based insecticides (20%) dramatically affect spore germination in *Asplenium scolopendrium* and *Athyrium filix-femina*, the process being reduced by 63.34–100% in the first species, respectively, 41.34–100% in the second. In *A.*

scolopendrium, in the variant exposed to 0.02% insecticide with acetamiprid, the gametophyte did not differentiate, the spores remaining ungerminated, while in *A. filix-femina*, after 6 weeks, the gametophyte was in the stage of three-dimensional cell masses and after 14 weeks in the cordate prothallium stage with archegonia [11].

The evaluation of glyphosate herbicide impact on *Blechnum appendiculatum, Macrothelypteris torresiana*, and *Thelypteris dentata* was made by applying it in concentrations of 0.33, 0.65, 2.72, and 10.89 g (active ingredient) l^{-1} at different stages of the life cycle: spore, gametophyte, and sporophyte [12]. The results indicated almost complete inhibition of the spore germination process, discoloration of the prothallic chlorocytes, as well as the chloroplastic tissues of the young sporophyte. The authors state that the herbicide has a negative impact on spore banks in the soil, resulting in a mortality of 31–50% in all green stages of their life cycle [12].

Also, the fern sporophytes are affected by pesticides. The relative growth rate, the amount of chlorophyll pigments, and the photosynthetic activity of *Azolla microphylla* aquatic fern were also significantly affected by treatment based on endosulfan, an insecticide applied in a concentration of 0–600 ppm. The authors consider the photosynthetic activity and the amount of chlorophyll pigments in order to explain how pesticides act on the photosynthetic mechanism in *Azolla* [13].

Saturn herbicide in *Azolla pinnata* caused the decrease of nitrogen, phosphorus, and potassium uptake in concentrations between 0 and 0.004 ppm applied for 5–25 days, while furadan insecticide in concentrations of 0.001 and 0.002 ppm stimulated uptake of nitrogen, phosphorus, and potassium after 20 and 25 incubation days on the culture medium [14]. Saturn also affected the fresh and dry mass of *Azolla pinnata.*

The paraquat herbicide applied in the culture medium of *Azolla microphylla*, in a concentration of 2–6 μM, resulted in overproduction of reactive oxygen species (ROS) which determined the induction of antioxidant enzyme activity: superoxide dismutase, catalase, guaiacol peroxidase, and peroxidase ascorbate, the species tolerating herbicide toxicity for 72 h [15]. In higher concentrations, this antioxidant defense mechanism is no longer working, and *Azolla* does not survive. Fragmentation and browning of fronds were also observed. In addition to the abovementioned changes, the presence of paraquat in the medium determines the decrease of both chlorophyll and protein amount [15].

The mature sporophyte of *Asplenium scolopendrium*, *Asplenium trichomanes*, *Athyrium filix-femina*, *Blechnum spicant*, *Dryopteris dilatata*, *Phegopteris connectilis*, *Polystichum* aculeatum, and *Woodsia ilvensis* species subjected *in situ* to treatment with asulam herbicide showed important damage, the maximum being recorded 1 year after application [16].

3. Influence of pesticides on histological and biochemical parameters in fish

As a result of penetration into the aquatic environment, pesticides affect a wide range of nontarget organisms such as invertebrates and fish [17]. Aquatic ecosystems are the "final destination" of pesticides used in agriculture [18]. Early life stages are the most sensitive. Many toxicity tests are focused on the study of hatching and the occurrence of deformities in hatched larvae, decreased mobility or even immobility, lack of coordination in swimming movements; for example, deltamethrin in concentrations >0.005 ppb decreases hatching rate in common carp [19]; bifenthrin in concentrations of 50–200 ppb determines the lack of coordination in swimming movements, malformations of axial development in *Danio rerio* species [20].

Acute exposure to pesticide resulted in reduced fish populations and increased mortality [21, 22]. Chronic exposure to small amounts of pesticide increased the incidence of disease, stress, and behavioral disorders [23]. Pesticide bioaccumulation causes a major danger—bioaccumulation factor of cypermethrin in fish is 1200× [24]. Acute tests of toxicity are used to evaluate pesticide toxicity and get rapid consistent results at least within certain limits, regarding the concentration-response relationship. These tests allow the determination of LC50 values for different periods of time, NOEC values, providing a very rich database on the toxic effect of many pesticides on different fish species.

Considering the large amount of data in the specialized studies on the effects of pesticides on fish, the data in this paper refer in particular to the pesticide groups on which we have conducted our own experiments. Many authors have established concentrations or lethal doses for different pesticides and species. For example, the toxicological values for chlorpyrifos are between 1.3 μg l^{-1} (96-h LC50) for *Lepomis macrochirus* and 2600 μg l^{-1} (72-h LC50 72) for *Gambusia affinis* [25]; LC_{50} in fish is less than 30 μg bifenthrin/l water [26, 27]; for most species, 96-h LC_{50} values for bifenthrin determined by the static method are between 2 and 5 μg l^{-1}, and between 0.5 and 4 μg l^{-1} by the continuous flow method [28]; propiconazole is moderate to slightly toxic for most aquatic organisms—96-h LC50 values for *Cyprinus carpio* 18.9 mg/l, *Oncorhynchus mykiss* 19 mg/l, and 21 mg/l for *Lepomis macrochirus* [29]; 48-h LC50 was 9 mg/l in *Carassius auratus*, 2.2 mg/l in *Oncorhynchus mykiss*, and 4 mg/l in *Cyprinus carpio*; the toxicity of chlorpyrifos in fish is generally between 0.01 and 1 mg/l (96-h LC50): 0.41 mg/l in *Oncorhynchus mykiss* and 0.015 mg/l in *Salmo gairdneri* [30].

As regards their action, pesticides block a certain metabolic process; the practical way in which this is done is sometimes difficult and in many cases is unknown or only partially clarified. The use of high sensitivity physiological indices is a tendency of recent years in aquatic toxicology. The most used physiological method is to determine pesticide action on fish breathing. Specialists can determine the concentrations in which breathing disorders begin to occur by recording the frequency of breathing movements and determining the oxygen consumption of fish kept in sublethal solutions of toxic substances (in state of rest or activity). The fish alters their energy metabolism in terms of spending a greater amount of energy to alleviate toxic stress [31], which results in improved oxygen use under hypoxia and anoxia conditions [32]. Determination and monitoring of oxygen consumption in aquatic organisms can be considered a better method to assess a substance toxicity than doing acute toxicological tests, because it has good results even in low concentrations of toxic substance [33]. One of the early symptoms of fish poisoning is breathing difficulties. Therefore, decreases in oxygen consumption in the gills were found by Cebrián et al. [34], even 24 h after exposure of *Procambarus clarkii* to chlorpyrifos, in the corresponding 96-h LC50 concentration. The frequency of opercular movements initially appears to support physiological activities in the polluted environment and is followed by the decrease in the breathing rate, which may be the result of gill diseases, as seen in the rainbow trout exposed to the action of fenvalerate and cypermethrin by Bradbury et al. [35].

Hematological parameters reflect fish state faster than determining other parameters because they change extremely rapidly under modified environmental conditions. For this reason, they are widely used to describe the state of fish health. Fish-nucleated red cells can be an ideal model to understand the harmful action of pesticides in different cell compartments; mitochondria and the nucleus of fish red blood cells are sensors in the programmed cell death mechanism [36].

Svobodova et al. reported significantly lower values of erythrocyte and hemoglobin in carp after acute exposure to deltamethrin (as a result of changes in hematopoiesis process), while the number of leukocytes did not change significantly [37]. Research carried out by Sopinska and Guz on permethrin poisoned carp revealed a decrease in the number of white blood cells, granulocytes in particular [38].

Anemia resulting from carp exposure to cypermethrin was also reported by Doruncu and Girgin [39].

Jayaprakash and Shettu evaluate the changes in some hematological parameters of the freshwater fish *Channa punctatus* exposed to sublethal concentrations of deltamethrin [40]. This study revealed: decrease in the hemoglobin content, total erythrocyte count, packed cell volume, mean corpuscular volume, mean corpuscular hemoglobin concentration, and a significant increase in the total leukocyte count.

Decrease in the number of erythrocytes in carp poisoned with pyrethroid insecticides (permethrin and cypermethrin), due to dysfunction of hematopoiesis [37, 39]. Exposure to Talstar EC 10 in a concentration of 57.5 $\mu g\ l^{-1}$ had no effect on the number of erytrocytes in common carp (*Cyprinus carpio*), with not significantly different values compared to those in the control group [41]. Atamanalp et al. [42] reported a significant increase in the number of erythrocytes in *Oncorhynchus mykiss* exposed to cypermethrin. Ahmad et al. observed significant decreases of hemoglobin in trout exposed to a sublethal dose of mancozeb, resulting in decreasing the amount of oxygen in the tissues and a decrease in the energy production of the animals [43]. Similarly, hemoglobin decreased percentually in *Heteropterus fossilis* after 30 days of exposure to deltamethrin [43].

Galloway and Handy studied the effect of organophosphorus pesticides (parathion, chlorpyrifos, malathion, and diazinon) on the immune system in invertebrates, fish, and other vertebrates [44]. Studies on fish have demonstrated an immunosuppressive effect to the action of these pesticides [45].

Physiological stress indicators, such as plasma cortisol and glucose level are commonly used to assess whether fish is stressed and whether the hypothalamo-pituitary-interrenal system works properly [46]. The intensive use of lactic acid in gill oxidation during stress caused by exposure to pesticides is mentioned in the literature [47] and is due to the increase in their energy demand [48] which is only covered by exogenous glucose. Glycogen mobilization is linked to the increase of energy demand; during pesticide-induced stress, a large amount of glycogen can be synthesized and stored in the muscles [49] which may later be in the form of glucose into blood tissues [50].

Due to the fact that fish live in a carbohydrate-deficient but protein-rich environment, they can use lipids and proteins as sources of energy more efficiently than carbohydrates [51]. Anaerobic glycolysis, recycling, and use of lactic acid preserve the energy potential of fish, thus ensuring better adaptation of fish to the polluted environment. Increases in the amount of energy (glucose) associated with lactate production have been reported in fish poisoned with pyrethroids [51].

High levels of lactic acid and strongly stimulated LDH activity in various tissues suggest the great importance of anaerobic glycolysis in adapting fish to stress caused by pyrethroid pesticides [52]. Velisek et al. examined the biochemical profile of carp (*Cyprinus carpio*) after 96 h exposure to bifenthrin in a concentration of 57.5 $\mu g\ l^{-1}$ and found significant increases in glucose level [41]. Bálint et al. observed an increase of blood glucose in carp after exposure to deltamethrin [53].

Velisek et al. showed an increase in blood glucose in trout and prussian carp as a result of metabolic stress induced by the action of Talstar 10 EC insecticide [41]. Jee et al. recorded an increase of blood glucose level in *Sebastes schlegeli* fish due to their exposure to cypermethrin (pyrethroid insecticide); researchers also reported a decrease in cholesterol and plasma proteins [54]. Datta and Kaviraj recorded increases in plasma glucose and decrease of liver glycogen in *Clarias gariepinus* exposed to a concentration of 0.005 mg cypermethrin/l [55]. Decrease in blood glucose levels following exposure to thyram was reported in *Salmo gairdneri* species by Van Leeuwen et al. [56].

Acetylcholinesterase is a highly used biomarker in aquatic ecotoxicology research [57], sensitive to low concentrations of organophosphates. Determination of acetylcholinesterase activity in erythrocytes is used to assess the degree of intoxication with acetylcholinesterase inhibitors. Recovery of acetylcholinesterase activity affected by exposure to organophosphorus pesticides can be carried out by dephosphorilation or AchE synthesis, which is a very slow process in fish; the disappearance of the inhibitory effect on acetylcholinesterase in *Gambusia affinis* exposed to oraganophosphorus compounds occurs within 45–60 days after exposure [58].

Costin et al. found a decrease of the specific activity of catalase, glutathione peroxidase, and glutathione reductase in prussian carps exposed to a sublethal concentration of deltamethrin (2 μg/l water), while glutathione-S-transferase activity increased [59]. Examination of the biochemical profile after 96 h of carp exposure to bifenthrin in a concentration of 57.5 μg l^{-1} resulted in significant increases in glucose, ammonia, aspartate aminotransferase, creatine kinase, and monocytes [41].

Bifenthrin inhibits the production of ATPase [60], which explains the stronger effect of bifenthrin on aquatic organisms compared to terrestrial ones (maintaining the critical concentration of ions against the concentration gradient, in the much diluted aquatic environment, requires intensive ionic transport processes, the necessary energy being supplied by ATPase; the decrease in production of this enzyme leads to the death of organisms). Li et al., following the study of propiconazole action on *Oncorhynchus mykiss*, suggested Na/K-ATPase activity in the fish brain as potential biomarkers for intoxication [61].

Fish gills are the main place of ion exchange with the environment and, at the same time, the main channel of pesticide penetration. They are in constant contact with water, and any change in its composition may affect the gills' permeability and their osmoregulatory functions. Cengiz found histopathological changes in the carp gills following acute exposure to deltamethrin in concentrations of 0.029 and 0.041 mg/l^{-1} (exfoliation, necrosis, edema, hyperplasia, fusion of secondary lamellae, etc.) [62]; similar changes caused by deltamethrin action were identified in *Gambusia affinis* by Cengiz and Unlu [63]. Costin et al. found morphological changes in the carp gills exposed to sublethal concentrations of deltamethrin (2 μg/l water), 48 h after the exposure to 2 μg deltamethrin/l water, which accentuated after 14 days of exposure (longest interval); the author reported hyperemia, fusion of secondary lamellae, epithelial layer damage, and chlorogenic cell proliferation [59].

Fusion of secondary lamellae as a result of exposure to pesticides appears to have a protective role in diminishing the affected gill surface; this response slows down the penetration of toxic and may result in fish choking [64, 65]. The toxic effect of pyrethroids on fish increases with increasing liposolubility [66]; because of their strong lipophilic character, the pyrethroids are well absorbed by the gills, even in low-pesticide waters, thus decreasing the availability of oxygen to the tissues of the internal organs [67].

Histological changes due to exposure to pesticides also occur in other organs. Melo et al. observed hepatic tissue alterations in *Rhamdia quelen* (loss of cell form, piknotic nuclei, necrosis, and increase of hepatocytes loaded with by gallbladder pigment) following the exposure of fish to organophosphates (0.01 ml/l Folidol for 4–72 h) [68]. Examination of liver tissue samples after 96 h of carp exposure to bifenthrin in concentration of 57.5 μg l^{-1} revealed hepatocyte degradation [41]. Cengiz and Unlu found histopathological changes caused by deltamethrin on the liver of *Gambusia affinis* [63]; cypermethrin causes hyperplasia and necrosis of hepatocytes in *Labeo rohita* species [69]. Ali et al. found DNA damage in erythrocytes and gill cells in *Channa punctatus* (Bloch), 5 days after exposure to chlorpyrifos in a concentration of 203 μg/l [70].

Much of the research on toxicological aspects of pesticide poisoning in fish refers to oxidative stress, causing the activity of antioxidant enzymes, proposed as biomarkers of pesticide intoxications [71–73]. It has been demonstrated that pyrethroid metabolism generates oxygen reactive species in poisoned fish [74]. The occurrence of oxidative stress in different fish tissues as a result of exposure to deltamethrin is supposed to be the main cause of product toxicity [75]. Chlorpyrifos toxicity is also manifested by the induction of oxidative stress [76]. It seems that oxidative stress contributes to the development and severity of syndromes in acute intoxication with these insecticides [77]. Mancozeb stimulated oxidative stress, while the amount of antioxidant serum enzymes decreased significantly; Fabra et al. demonstrated that this fungicide caused significant biochemical changes in cell membranes [78].

Exposure to pesticides may bring about behavioral changes such as: hyperactivity, loss of balance, and impossibility of maintaining the normal position in trout exposed to cypermethrin [79]. Symptoms of fish exposed to pyrethroid pesticides are: swimming near water surface, hyperactivity, loss of balance, increase of branchial mucus secretion, etc. Levine et al. found a decrease of swimming intensity in *Danio rerio* due to exposure to chlorpyrifos in a concentration of only 100 mg/l [80].

Toxicity of different pesticides on fish and other aquatic organisms can be influenced by various factors such as: pH, temperature, dissolved oxygen, water hardness, etc. Mauck et al. found that bifenthrin is more toxic at low temperatures, and its toxicity is only slightly influenced by water hardness [81]. Macek et al. recorded 96-h LC50 values of 7.1, 15, and 51 $\mu g\ l^{-1}$ in *Oncorhynchus mykiss* at temperatures of 12.7, 7.2, and 1.6°C [82].

Although the data on histochemical, biochemical, hematological, and behavioral changes in poisoned fish are quite numerous, they are little related to *Carassius gibelio, Perca fluviatilis*, and *Alburnus alburnus* species which are not commonly used in toxicological tests. In toxicology research, the emphasis is placed on biochemical, chemical, structural, and ultrastructural changes, disregarding the physiological processes which are part of the primary response of organisms to stressors.

As a contribution to the study of pesticide effect on fish, we will briefly present some of our results for these species exposed to the action of six pesticides—3 insecticides (Talstar One—in concentrations of 0.05, 0.1, 0.2 and 0.4 µl bifenthrin/l water, Reldan 40 EC—0.4, 0.8, 1.6, and 3.2 mg chlorpyrifos/l water, and Actara 25 WG—0.016, 0.032, 0.064 and 0.128 mg thiamethoxam/water) and 3 fungicides (Tilt 250 EC—1.06, 2.12, 4.24, and 3.2 mg chlorpyrifos/l water, Dithane M-45—2, 4, 8, and 16 mg mancozeb/l water, and Tiradin 70 PUS—0.007, 0.014, 0.028, and 0.56 mg thiram/l water).

The analyzed pesticides reduce energy metabolism and breathing rhythm in crucian carp, bass, and bleak [83–92]. In many of the variants, the immediate response to the toxic action is a breathing rate stimulation of variable duration, in inverse ratio to the pesticide concentration, with high decrease in oxygen consumption, followed by the "stabilization phase" to the new conditions, with small fluctuations of the mentioned indices around an average value [85–92]. The advantage of determining energy metabolism and breathing rate is the rapidity of this response to the action of the stressor (pesticide) on the one hand and their noninvasive character on the other.

The number of erythrocytes increases after exposure to Reldan 40 EC, Tilt 250 EC, and Tiradin 70 PUS [86–89, 92]; decreases of this hematologic parameter are recorded under the action of Actara 25 WG and Talstar One [84, 90, 92]. Blood glucose levels in fish increase for 2 weeks under the action of Talstar One, Actara 25 WG, and Tilt 250 EC [84–86, 90, 92]. Dithane M-45 fungicide and Reldan 40 EC insecticide have a hypoglycemic effect [83, 86, 92]. Tiradin does not change blood glucose level in crucian carp and bleak intoxicated with a concentration of 0.02 ml/l for 2 weeks [91, 92].

Behavioral changes occur in crucian carp and perch exposed to pesticides: hyperactivity in the first 24–48 h, disordered movements, frequent rising at the surface of water in the first hours after exposure to toxic, and apathy under the action of insecticides and fungicides [85–92]. Increases in mucus production occur especially in crucian carp under the action of fungicides [83, 87–89, 91, 92].

There are different results as regards the sensitivity of the species: the prussian carp was more sensitive under the action of Reldan 40 EC and Actara 25 WG [86, 92] as well as Tiradin 70 PUS and Tilt 250 EC fungicides [87, 88, 91, 92]; the bleak was more sensitive to Talstar One and Dithane M-45 [92]. In some experimental variants, there were no significant differences in the reactivity of the species under the action of the six pesticides.

Among the physiological parameters, blood glucose values and the number of red blood cells give a clear signal of fish stress, which is why we recommend using these parameters as biomarkers for pesticide-induced toxic stress. The study of oxygen consumption and breathing rate combined with the number of erythrocytes and blood glucose as well as the main behavioral changes allow the formation of a symptomatological picture of fish poisoned with pesticides.

4. Influence of pesticides on histological and biochemical parameters in amphibians

The impact of pollutants on humans in their environment is particularly complex and explains the lack of action or late and often confusing reactions regarding the protection measures. The degree of exposure depends on the simultaneous presence of some essential factors (nature and concentration of pollutants) and modifiers that ultimately influence the absorbed quantity.

Amphibians are some of the most representative species of vertebrates in aquatic and agricultural ecosystems, because they are the natural enemies of many pests of crop plants. Due to their high sensitivity to the changes occurring in their natural environment and because their larval development occurs exclusively in the aquatic environment, amphibians have been considered bioindicators for the aquatic and agricultural ecosystems [93–96] and have been used as test specimens to study the action of different chemicals in these ecosystems.

Bridges [97] conducted numerous studies on the effects of long-term exposure to pesticides. The presence of contaminants in the environment may alter the predator-prey interaction among aquatic species by changing the predator or prey levels of activity or behavioral change of the predator. The effect of predator-prey meeting may be dependent on the fact that both species are exposed simultaneously to a contaminant/pollutant or exposure occurs only to one of the two species.

Due to the short period of exposure to nonpersistent pollutants, it is important to examine the long-term effects of short-term exposure on amphibians and the sensitivities at a certain stage in their development cycle. Any delay in metamorphosis or any decrease in size during metamorphosis may have an impact on the evolution of the amphibian populations, leading to its decline or local extinction.

Sampath et al. [98] studied the toxic effect of two pesticides (carbaryl and methyl-parathion) on the excretory system of *Rana tigrina* tadpoles and showed an increase in the rate of N-urea and NH3-N elimination in intoxicated specimens depending on the concentration of pesticide. Short-term exposure (96 h) to action in low doses of endosulfan organochlorine insecticide caused changes in the growth of *Litoria freycinetti* tadpoles either immediately or on a long term [99]. Tadpoles exposed to toxic action grew more slowly, and those who survived were more easily captured by predators in their natural environment.

Tomizawa and Casida [100, 101] demonstrated that neonicotinoid insecticides acted selectively on the nicotine-acetylcholine receptor in insects and mammals, making them the safest insecticides (as a mode of action). They described the mechanism of selective toxicity of neonicotinoid insecticides.

Gendron et al. [102] published the conclusions of a study on the effects of the leopard frogs' exposure to a mixture of pesticides. The hypothesis shows that the exposure of *Rana pipiens* leopard frogs to agricultural pesticides can affect the dynamics of the parasitic worm infections, *Rhabdias ranae*. The pesticide treatment did not influence the growth of the parasites, the results indicating that they got matured and reproduced earlier in frogs exposed to pesticides, compared to control specimens. Such alternations in developmental cycle characteristics that increase the transmission of parasites may lead to an increase in virulence. The results contribute to further discussion on the role that anthropogenic factors might have in the gradual death of amphibians due to complications arising from the disease observed in different parts of the world.

Greulich and Pflugmacher [103] suggested a number of factors for the recent decrease in amphibian populations, one of them being exposure to pesticides. Specialists studied the absorption and effects of the pyrethroid cypermethrin insecticide in relevant concentrations from the environment for two different amphibian species, *Bombina variegata* and *Rana arvalis*, with the observation that cypermethrin absorption caused deformities, abnormal behaviour, and mortality. These changes depended on the dose of pesticide.

Some studies show that glyphosate has a harmful effect on the environment, especially on amphibians. Howe et al. [104] studied the effect of glyphosate on four North American amphibian species: *Rana clamitans*, *R. pipiens, R. sylvatica*, and *Bufo americanus*, and observed the following aspects: decrease of the anteroposterior diameter of the body, increased time for metamorphosis, tail malformations in tadpoles, and gonadal disorders. These effects may partly appear as a result of changes in the hormone level. There was a high level of transcription for ARNm segments encoding β-receptor synthesis for thyroid hormones, following exposure of individuals to solutions containing glyphosate.

Chen et al. [105] conducted studies on the effect of glyphosate on fauna in wetlands, having the zooplankton, *Simocephalus vetulus*, and *Rana pipiens* tadpoles as research subject. Significant effects of the pesticide action on the two species were determined at concentrations lower than those expected in the environment. Increased pH values (7.5) also showed the toxic effects of the pesticide on the two species under study, although the reproduction rate for *S. vetulus* improved to a pH level above 5.5 in the absence of the pesticide. The stress caused by the lack of food, associated with a pH of 5.5, decreased the survival rate of *S. vetulus.*

Relya [106] studied how four commercial compounds (diazinon, carbaryl, malathion, and glyphosate) affected the survival and larval growth rate of five species of amphibians (*Rana pipiens*, *R. clamitans*, *R. catesbeiana*, *Bufo americanus*, and *Hyla versicolo*). The combination of pesticides has occasionally resulted in lower survival rates and development than those determined by each pesticide, but never lower than those caused by the worse of the two. This suggests that the combination of the four pesticides had the same effect as the estimated total concentration of pesticides on the ecosystem.

The toxicity of pesticides in general and their genotoxicity on nontarget organisms in particular was of special interest to researchers. Shaolong et al. [107] studied the toxicity and genotoxicity of two pesticides (imidacloprid and RH-5849), used in China since 1992, for two species of amphibians. RH-5849 insecticide did not prove to be toxic to tadpoles even if they were kept for 96 h in a saturated solution. Two techniques were used to study the genotoxicity of the two insecticides: micronucleus test and single cell gel electrophoresis. These tests showed significant changes in the DNA of amphibian erythrocytes.

Relya [108] studied six North American species of amphibian larvae (*Rana sylvatica*, *R. pipiens*, *R. clamitans*, *R. catesbeiana*, *Bufo americanus*, and *Hyla versicolor*) to investigate the long-term effects of Roundup, many of the existing studies focusing on short-term effects (1–4 days). He also studied the effects of Roundup's association with other sources of stress, such as predators, on the tadpoles' survival rate during 16 days with and without chemical signals from predatory salamanders (*Notophthalmus viridescens*).

The values range from 0.55 to 2.52 mg of active ingredient (IA)/L, considerably lower than those used in previous studies. Stress increased by predators made Roundup two times more toxic to one of the six species (*R. sylvatica*). This finding suggests that the synergistic action of pesticides and predators may be a general phenomenon for amphibians (the range of pesticides can be very wide). Based on this research, pesticides such as Roundup certainly play a significant role in the decline of amphibian populations around the globe.

Relyea et al. [109] also studied the effect of pesticides (glyphosate and malation) on the natural environment of amphibians in the presence of zooplankton and phytoplankton (algae). Three species of amphibian larvae (*Hyla versicolor, Bufo americanus,* and *Rana pipiens*) were studied and combined with their predators (no predators with *Notophthalmus viridescens* and *Dytiscus sp*. Larvae) and pesticides (no pesticides, with malathion insecticide and Roundup herbicide). Roundup proved to have a negative effect on tadpoles, reducing their biomass by 40%, with no indirect effects on the amphibian community through predators or abundance of algae. Malathion in a concentration of 0.3 mg/l caused the number of tadpoles to decrease. This insecticide associated with the tritone in the amphibian environment does not have significant effects; the presence of beetle larvae associated with malathion has a positive effect, which determines the increase in the survival rate of the tadpoles in parallel with the decrease in the number of predators. While high concentrations of malathion can cause the death of amphibian larvae, the small concentrations may have positive effects as beetle larvae, and the predators of amphibian larvae are killed. These data lead us to the conclusion that pesticides can have direct and indirect effects on natural amphibian communities.

Toxicological investigations on amphibian larvae have mainly focused on the effects of heavy metals and pesticides on their growth, development, and behaviour, and only a few data refer to the bioaccumulation of toxins under natural conditions. Hofer et al. [110] studied the accumulation of inorganic (Pb and Cd) and organic (organochlorine and polyaromatic hydrocarbon pesticides) toxic in the body of *Rana temporaria* tadpoles at various altitudes in the Alps. They found an increase in the accumulation of Pb and Cd in the body of tadpoles in low pH waters, with a high concentration of metals. The amount of organochlorine substances was relatively low due to the age of the tadpoles (about 2 months) and the absence of lipid deposits necessary for the absorption of these substances.

Seifert and Stollberg [111] investigated the interaction of the neonicotinoid imidacloprid insecticide with the nicotine-acetylcholine receptor (nAcChR) in the frog embryonic muscle cells. The response of the muscle cell to the action of acetylcholine, nicotine, and imidacloprid was recorded as a contraction. The contractions of acetylcholine or nicotine are inhibited by α-bungarotoxin. Imidacloprid does not lead to the contraction of the muscle cell, but it can attenuate the contractions produced by acetylcholine or nicotine. They have found that imidacloprid is an antagonist of nAcChR receptor in the muscle cell and an agonist in its toxic action on insect nerve receptors.

Honda et al. [112] studied how to activate or inactivate 11 nicotinic insecticides in a two-step system of metabolic coupling and receptor binding. The authors incubated neonicotinoid insecticides with CYP3A4 and NADPH or AOX on a cosubstrate of N-methyl-nicotinamide for metabolism. They used ketoconazole or menadione for inhibition of subsequent conversions. They also used *Drosophila* nAChR receptor and [^{3}H] imidacloprid or α4β2 and [^{3}H](−)-nicotine receptor to determine changes in the action of neonicotinoid insecticides. *Drosophila* nAChR receptor coupled with CYP3A4 system activates imidacloprid and thiamethoxam, while the other nine insecticides do not undergo any change in their toxic potential. AOX system coupled with *Drosophila* nAChR receptor strongly inactivates clothianidin, dinotefuran, imidacloprid, desmethyl-thiamethoxam, and thiamethoxam has a low inactivation effect on nitenpiram and nithiazines but has no effect on the other four insecticides studied.

Neurotoxicity of pesticides has been studied by many researchers on the sciatic nerve of *Rana ridibunda*. Zafeiridou et al. [113] studied the influence of three herbicides: acetochlor, 2,4,5-trichlorophenoxyacetic acid (2,4,5-T), and 2,4-dichlorophenoxyacetic acid, on the value of the action potential of the sciatic nerve in *Rana ridibunda*. Since the action of 2,4-D is pH-dependent, the toxicity of the three chemical compounds has been studied at a pH of 3.3. It has been found that a low pH increases the toxicity of 2,4,5-T and 2,4-D pesticides and decreases acetochlor toxicity.

Casco et al. [114] demonstrated high toxicity of cypermethrin in *Bufo arenarum* larvae. They showed that low doses of cypermethrin, lower than those used in agriculture, can cause massive apoptosis in the central nervous system of the tadpoles through a neurotoxic mechanism.

In 2006, Relya [115] published an article on the effects of pesticides, pH, and stress due to predation on amphibians. It was shown that pH and stress due to predation and a single application of the insecticide (carbaryl) affected the survival and growth of *Rana catesbein* larvae and *Rana clamitans* green frogs under natural conditions. Decreasing the pH had no effect on the survival rate but caused a faster growth of tadpoles. Low concentrations of carbaryl did not affect the two species, but higher concentrations resulted in a lower survival rate and higher increases in individuals of *Rana catesbein* species. The stress caused by predation and low pH did not increase herbicide lethality due to its rapid rate of degradation under natural environmental conditions. Even if stress, pH, or predators could make carbaryl more lethal under laboratory conditions, after repeated insecticide applications, these stress factors did not interact under natural conditions even if a single carbaryl application was used.

The application of auxin herbicides of 2,4-dichlorophenoxyacetic acid (2,4-D) type in deciduous forests is mainly used in reafforested areas. The absorption of this herbicide in the target plants is reduced due to its low solubility in water. Because of this, the acid is esterified to increase its liposolubility and to diffuse into the plant-bearing vessels. But the esterification of auxin herbicides increases their absorption in other organisms, such as amphibians [116].

Numerous researches have shown that the gelatinous shell of amphibian embryos has a protective role against various environmental pollutants. Edginton et al. [117] studied the rate of absorption, metabolism, and excretion of butoxyethyl ester for 2,4-dichlorophenoxyacetic acid (2,4-D BEE) in *Xenopus laevis* embryos. They found that embryos with a gelatinous coating had a low absorption of toxic compared to embryos without the coating after 8 h from exposure. Metabolism of 2,4-D BEE lasted between 35 and 42 min, and the accumulation of radioactive residue was 35% located in the gelatinous shell and 65% in the embryo.

Vonesh and Buck [118] have shown that pesticides influence the process of laying eggs in amphibians. Four experimental batches were used to examine the effect of the pesticide marketed under the name of Sevin® and the active substance, carbaryl, on the method of laying eggs in *Hyla chrysoscelis* species. Their results have shown that unpolluted ponds are preferred by amphibians for laying eggs, while the presence of Sevin® and carbaryl in the environment reduces the process, while volatile chemicals have no influence at all.

Gurushankara et al. [119] investigated the effect of malathion in various concentrations on the survival, growth, and food consumption in *Limnonectus limnocharis*. Exposure of tadpoles to malathion changed all parameters. Thus, there was a decrease in survival rate, body size in parallel with the increase in malathion concentration in the environment as well as a reduction in the amount of food consumed.

Bioaccumulation of malathion and its impact on *Ambystoma tigrinum* was studied by Henson-Ramsey et al. [120]. The toxic was administered by two methods: soil contamination with malathion in a concentration of 50 and 100 μg/cm^2 and food—the administration of earthworms that lived on a malathion-contaminated soil in a concentration of 200 μg/cm^2.

The amount of xenobiotic substances was determined by gas chromatography in liver, muscle, skin, and bones. Malathion was higher in muscle, skin, and bones after 1 day of exposure and in viscera, after 2 days of exposure. To determine the biological impact of malathion on *Ambystoma tigrinum*, cholinesterase activity was measured in the brain, with a reduction in its activity. Animals exposed to toxic action did not show clinical signs of toxicosis.

Aquatic organisms are exposed to repeated pesticide applications over time. Repeated and long-term (79 days) administration of low concentrations of malathion (10–250 μg/l)—the most widely used insecticide, affects the entire aquatic community of zooplankton, phytoplankton, periphyton, and amphibian larvae [121]. This caused the decline of zooplankton followed by an increase in the biomass of the phytoplankton and the decline of periphyton after repeated treatments. The decline of the periphyton has had little influence on the development of *Rana sylvatica* species because it has a short time of metamorphosis, but has strongly influenced *Rana pipiens* larvae which have a longer time of metamorphosis with a high mortality level. In conclusion, malathion did not directly kill amphibian larvae but caused changes in the trophic chain and caused their mortality in an indirect way.

Toxicity caused by endosulfan (an organochlorine chemical compound used as an insecticide in vegetable, fruit, cotton, and coffee) was studied on *Bombina orientalis*, by Kang et al. [122]. They found a decrease in the survival rate of the tadpoles in parallel with an increase in the toxic concentration. The surviving tadpoles showed morphological changes consisting of tail dysplasia, presence of vesicles in the pectoral and ventral side, tail bending, and cephalic dysplasia; the higher the toxic concentration, the more significant the morphological changes.

Similar changes have also been reported in intoxication of larvae and *Bombina orientalis* tadpoles with molinate, a thiocarbamate used as an herbicide [123]. The change in survival rate was not significant until the blastula stage. A significant decrease of this parameter was observed in the tadpole stage even in low doses of molinate. The surviving tadpoles showed many malformations: trunk and tail bending, tail and eye dysplasia, and cephalic dysplasia. All these data suggest that the molinate herbicide influenced the larval development of this species.

Relyea and Jones [124] set LC50 after 96 h for glyphosate (Roundup) for various North American amphibian species. There were used nine species of larval anura Groser 25 belonging to *Raniidae*, *Bufonidae*, and *Hylidae* families, as well as four larval urodeles from two families: *Salamandridae* and *Ambystomatidae*. LC50 was between 0.8 and 2 mg acid/l for the nine anura species after 96 h and ranged from 2.7 to 3.2 mg acid/l for the four larval urodele. This work brings new data on the sensitivity of amphibians to the action of glyphosate.

Endosulfan insecticide appears to be highly toxic in low doses for amphibian populations and also exhibits long-term toxic effects. Toxicological studies conducted by Jones et al. [125] established LC50 in 4 days (LC504-d) for endosulfan insecticide on tadpoles of nine amphibian species belonging to three families (*Bufonidae*, *Hylidae*, and *Ranidae*). LC504-d value was between 1.3 and 120 ppb, which places it in the category of highly toxic pesticides. At the end of the treatment, experimental animals were kept in clean water for 4 more days. There has been a significant increase in mortality depending on the family. Bufonides were the least sensitive, hylides were moderately sensitive, and ranidae were the most sensitive.

Atrazine is a widely used herbicide for weed control in corn crops. Its residues could be detected in the soil 1 year after application. Numerous studies have tracked the effect of atrazine on embryonic development of amphibians in the environment and reported its ability to induce an enzyme—aromatase, which appears to convert testosterone to estrogen [126–132].

Some studies followed the effect of atrazine on juvenile and adult amphibians. Storrs et al. [133] followed the effect of atrazine administered directly on the soil on the behaviour of *Bufo americanus* species. They also studied various ways to absorb, accumulate, and eliminate this toxic from the body using ^{14}C-labeled atrazine. Their results have shown that amphibians do not avoid atrazine-enriched soils. This pesticide is rapidly absorbed through the skin, and accumulated in the intestine, especially in the gallbladder.

Small quantities of herbicides used in agriculture often affect the surface of water and can become a stress factor for aquatic organisms. Williams and Semlitsch [134] studied the effect of 4 herbicides (the active substance atrazine, S-metolachlor and glyphosate) on the amphibian larvae of 3 species: *Bufo americanus*, *Pseudacris triseriata* and *Hyla versicolor*. The glyphosate under the trade name Roundup WeatherMax (containing 572 ppb equivalent glyphosate acid) determines the increase in mortality of *Pseudacris triseriata* tadpoles, while Roundup OriginalMax (containing 572 ppb equivalent glyphosate acid) determines the extension of the larval period in *Bufo americanus* species.

Chronic exposure to atrazine and S-metolachlor has no side effects on survival, metamorphosis, or larval period in the organisms. These results show the importance of adjuvants in the preparation of pesticides and their different effects on organisms even in the case of similar products.

Budischak et al. [135] followed how some parasites of amphibians can influence the toxicity of some pesticides. They used two batches of *Rana palustris*: tadpoles infected with *Echinostoma trivolvis* trematode, then intoxicated with malathion and uninfected tadpoles, and intoxicated with the same dose of malathion. It has been shown that the parasite does not influence susceptibility to pesticide action.

The influence of six pesticides (Reldan 40EC, Actara 25WG, Tilt 250EC, Champion 50WG, Fusilade Forte, and Dual Gold 960EC) on physiological indices of *Pelophylax ridibundus* was studied by Paunescu et al. [136]. The chronic pesticides exposure can lead to alteration of various indices (increased hepatosomatic index, decrease in erythrocytes number, leucopenia, hyperglycaemia, and decrease in cholesterol levels), as well as to hepatic lesion.

The fact that some species of amphibians survive in agroecosystems is due to their special plasticity. Thus, irrigation canals represent new habitats that amphibians colonize, and in the absence of predators, which are less mobile (fish), they become metapopulations of altered landscapes [137–139].

5. Conclusions

As nontarget species for pesticides in the environment, ferns can be used in their gametophyte stage, young or mature sporophyte in different biotests to evaluate the risk associated with these substances. The biochemical, hemathological, and histopathological changes recorded in both fish and amphibians can be considered biomarkers of pesticide pollution.

Acknowledgements

This work was supported by a grant of the Romanian Ministery of Research and Innovation, CCCDI-UEFISCDI, project number PN-III-P1-1.2-PCCDI-2017-0332"/ Project 3 "Development of plant extracts and innovative phytosynthesized nano-structured mixtures with phytotherapeutical applications aimed at eliminating the biocenotic stress in horticultural crops", within PNCDI III.

Author details

Liliana Cristina Soare*, Alina Păunescu and Ponepal Cristina Maria
Department of Natural Sciences, Research Centre for the Nature Protection, University of Pitești, Pitești, Romania

*Address all correspondence to: soleil_cri@yahoo.com

References

[1] Sharpe JM, Mehltretter K, Walker LR. Ecological importance of ferns. In: Mehltretter K, Walker LR, Sharpe JM, editors. Fern Ecology. Cambridge: Cambridge University Press; 2010. pp. 1-21. DOI: 10.1017/CBO9780511844898.002

[2] Smith AR, Pryer KM, Schuettpelz E, Korall P, Schneider H, Wolf PG. Fern classification. In: Ranker TA, Haufler CH, editors. Biology and Evolution of Ferns and Lycophytes. Cambridge: Cambridge University Press; 2008. pp. 417-467

[3] Rodríguez-Romero AJ, Sedeño-Díaz JE, López-López E, Esteban M, Quintanilla LG, Catalá M. Recent advances in the use of mitochondrial activity of fern spores for the evaluation of acute toxicity. In: Fernándes H, editor. Current Advances in Fern Research. Cham: Springer International Publishing; 2018. pp. 481-498. DOI: 10.1007/978-3-319-75103-0

[4] García-Cortés H, Catalá M, Rodríguez Gil JL. Update on the assessment of chronic phytotoxicity using fern spore biomarkers. In: Fernándes H, editor. Current Advances in Fern Research. Cham: Springer International Publishing; 2018. pp. 499-515. DOI: 10.1007/978-3-319-75103-0

[5] Ehrendorfer F. Pteridophyta. In: Sitte P, Ziegler H, Ehrendorfer F, Bresinsky A, editors. Strasburger—Lehrbuch der Botanik. 34 Auflage ed. Berlin: Spektrum Akademischer Verlag Heidelberg; 1999. pp. 652-684

[6] Banks JA. Gametophyte development in ferns. Annual Review of Plant Physiology and Plant Molecular Biology. 1999;**50**:163-186. DOI: 1040-2519/99/0601-0163$08.00

[7] Catalá M, Esteban M, Rodríguez-Gil JL, Quintanilla LG. Development of a naturally miniaturised testing method based on the mitochondrial activity of fern spores: A new higher plant bioassay. Chemosphere. 2009;**77**:983-988. DOI: 10.1007/s10646-011-0790-2

[8] Droste A, Brizola Cassanego MB, Windisch PG. Germination and sporophytic development of *Regnellidium diphyllum* Lindm. (Marsileaceae) in the presence of a glyphosate-based herbicide. Revista Brasileira de Biociências. 2010;**8**:174-178

[9] Soare LC, Dobrescu CM, Popescu M, Boeru AG. The effects of some pesticides on spore germination and gametophyte differentiation in *Athyrium filix-femina* (L.) Roth. and *Polypodium vulgare* L. Notulae Botanicae Horti Agrobotanici Cluj-Napoca. 2013;**41**: 458-462. DOI: 10.15835/nbha4129088

[10] Soare LC, Dobrescu CM, Popescu AG. The influence of a copper-containing funcicide on the gametophyte of some non-target pteridophyte species. Acta Horti Botanici Bucurestiensis. 2013;**40**:49-56. DOI: 10.2478/ahbb-2013-0006

[11] Soare LC, Dobrescu CM, Burtescu L. Research on the influence of two insecticides on the gametophyte of some leptosporangiate pteridophytes. Analele Ştiinţifice ale Universităţii "Al. I. Cuza" Iaşi, Biologie Vegetală. 2013;**59**:5-12

[12] Aguilar-Dorantes K, Mehltreter K, Mata-Rosas M, Vibrans H, Esqueda-Esquivel V. Glyphosate susceptibility of different life stages of three fern species. American Fern Journal. 2015;**105**:131-144. DOI: 10.1640/0002-8444-105.3.131

[13] Raja W, Rathaur P, John SA, Ramteke PW. Endosulfan induced changes in growth rate, pigment composition and photosynthetic activity of mosquito fern *Azolla microphylla*.

Journal of Stress Physiology & Biochemistry. 2012;**8**:98-109

[14] El-Shahate RM, El-Araby MMI, Eweda EW, El-Berashi MN. Evaluation of the effect of three different pesticides on *Azolla pinnata* growth and NPK uptake. Journal of American Science. 2011;7:1020-1031

[15] Sooda A, Pabbib S, Uniyala PL. Effects of paraquat on lipid peroxidation and antioxidant enzymes in aquatic fern *Azolla microphylla*. Russian Journal of Plant Physiology. 2011;**58**:667-673. DOI: 10.1134/S1021443711040170

[16] Rowntree JK, Sheffield E. Effects of asulam spraying on non-target ferns. Canadian Journal of Botany. 2005;**83**:1622-1629. DOI: 10.1139/b05-127

[17] Otludil B, Cengiz EI, Yildirim MZ, Unver O, Unlu E. The effect of endosulfan on the great ramshorn snail *Planorbarius corneus* (Gastropoda: Pulmonata): A histological study. Chemosphere. 2004;**56**:707-716. DOI: 10.1016/j.chemosphere.2004.04.027

[18] Wang C, Lu G, Cui J, Wang P. Sub-lethal effects of pesticide mixtures on selected biomarkers of *Carassius auratus*. Environmental Toxicology and Pharmacology. 2009;**28**:414-419. DOI: 10.1016/j.etap.2009.07.005

[19] Köprücü K, Aydın R. The toxic effects of pyrethroid deltamethrin on the common carp (*Cyprinus carpio* L.) embryos and larvae. Pesticide Biochemistry and Physiology. 2004;**80**:47-53. DOI: 10.1016/pestbp.2004.05.004

[20] Jin M, Zhang X, Wang L, Huang C, Zhang Y, Zhao M. Developmental toxicity of bifenthrin in embryo-larval stages of zebrafish. Aquatic Toxicology. 2009;**95**:347-354. DOI: 10.1016/j.aquatox.2009.10.003

[21] Fulton MH, Moore DW, Wirth EF, Chandler GT, Key PB, Daugomah JW, et al. Assessment of risk reduction strategies for the management of agricultural nonpoint source pesticide runoff in estuarine ecosystems. Toxicology and Industrial Health. 1999;**15**:201-214. DOI: 10.1177/074823379901500118

[22] Grant IF. Insecticides for tsetse and trypanosomiasis control: Is the environmental risk acceptable? Trends in Parasitology. 2001;**17**(1):10-14. DOI: 10.1016/S1471-4922(00)01848-1

[23] Austin B. The effects of pollution on fish health. Journal of Applied Microbiology Symposium Supplement. 1999;**85**:234S-242S. DOI: 10.1111/j.1365-2672.1998.tb05303.x

[24] U.S. Environmental Protection Agency. Pesticide Fact Sheet Number 199: Cypermethrin. Washington, D.C.: Office of Pesticides and Toxic Substances; 1989

[25] Davey RB, Meisch MV, Carter FL. Toxicity of five ricefield pesticides to the mosquitofish, *Gambusia affinis*, and green sunfish, *Lepomis cyanellus*, under laboratory and field condition in Arkansas. Environmental Entomology. 1976;**5**:1053-1056. DOI: 10.1093/ee/5.6.1053

[26] Velisek J, Dobsikova R, Svobodova Z, Modra H, Luskova V. Effect of deltamethrin on the biochemical profile of common carp (*Cyprinus carpio* L.). Bulletin of Environmental Contamination and Toxicology. 2006;**76**:992-998. DOI: 10.1007/s00128-006-1016-9

[27] Velisek J, Jurcikova J, Dobsikova R, Svobodova Z, Piackova V, Machova J, et al. Effects of deltamethrin on rainbow trout (*Oncorhynchus mykiss*). Environmental Toxicology and Pharmacology. 2007;**23**:297-301. DOI: 10.1016/j.etap.2006.11.006

[28] Clark JR, Partick JM, Middaugh DP Jr, Moore JC. Relative sensitivity of six estuarine fishes to carbophenothion, chlorpyrifos, and fenvalerate. Ecotoxicology and Environmental Safety. 1985;**10**:382-390. DOI: 10.1016/0147-6513(85)90083-1

[29] Envirofacts. Cyproconazole The Active Ingredient in Alto® One of the Active Ingredients in Quadris Xtra®. Vol. 2. Greensboro, NC: Syngenta Crop Protection Inc; 2005. pp. 7419-8300

[30] Van Wijngaarden R, Leeuwangh P, Lucassen WG, Romijn K, Ronday R, Velde R, et al. Acute toxicity of chlorpyrifos to fish, a newt, and aquatic invertebrates. Bulletin of Environmental Contamination and Toxicology. 1993;**51**:706-712. DOI: 10.1007/BF00201650

[31] Ferrando MD, Andreu-Moliner E. Lindane induced changes in carbohydrate metabolism in *Anguilla anguilla*. Comparative Biochemistry and Physiology. 1992;**74B**:750-799. DOI: 10.1016/0742-8413(92)90300-V

[32] McCormic SD, Moyes SD, Ballantynes JS. Influence of salinity on the energitics of gill and kidney of Atlantic salmon (*Salmo salar*). Fish Physiology and Biochemistry. 1989;**6**: 243-254. DOI: 10.1007/BFO1875027

[33] Bakthavathasalam R. Effects of lindane and carbofuran on the lactic acid levels of the fish, *Anabas testudineus* at submerged condition and on exposure to air. Environment and Ecology. 1988;**6**:32-37

[34] Cebrián C, Andreu-Moliner ES, Fernández-Casalderrey AS, Feraldo MD. Acute toxicity and oxygene consumption in the gills of *Procambarus clarkia* in relation to chlorpyrifos exposure. Bulletin of Environmental Contamination and Toxicology. 1992;**49**:145-149. DOI: 10.1007/BF00193353

[35] Bradbury SP, Coats JR, McKim JM. Toxicokinetics of fenvalerate in rainbow trout *Salmo gairdneri*. Environmental Toxicology and Chemistry. 1986;**5**:567-576. DOI: 10.1002/etc.5620050609

[36] Bratosin D, Estaquier J, Slomianny C, Tissier JP, Quatannes B, Bulai T, et al. On the evolution of erythrocyte programmed cell death: Apoptosis of *Rana esculenta* nucleated red blood cells involves cysteine proteinase activation and mithochondrion permeabilization. Biochimie. 2004;**86**(3):183-193. DOI: 10.1016/j.biochi.2004.03.003

[37] Svobodova Z, Luskova V, Drastichova J, Svoboda M, Zlabek V. Effect of deltamethrin on haematological indices of common carp (*Cyprinus carpio* L.). Acta Veterinaria Brno. 2003;**72**:79-85. DOI: 10.2754/avb200372010079

[38] Sopinska A, Guz L. Influence of permethrin on phagocytic activity of carp. Medycyna Weterynaryjn. 1998;**54**:126-128

[39] Dorucu M, Girgin A. The effects of cypermethrin on some haematological parameters of *Cyprinus carpio*. Aquaculture International. 2001;**9**: 183-187. DOI: 10.1023/A:1014297528028

[40] Jayaprakash C, Shettu N. Changes in the hematology of the freshwater fish, *Channa punctatus* (Bloch) exposed to the toxicity of deltamethrin. Journal of Chemical and Pharmaceutical Research. 2013;**5**(6):178-183

[41] Velisek J, Svobodova ĆZ, Machova ĆJ. Effects of bifenthrin on some haematological, biochemical and histopathological parameters of common carp (*Cyprinus carpio* L.). Fish Physiology and Biochemistry. 2009;**35**:583-590. DOI: 10.1007/s10695-008-9258-6

[42] Atamanalp M, Yanýk T, Haliloûlu I, Aras MS. Alterations

in the hematological parameters of rainbow trout (*Oncorhynchus mykiss*) exposed to cypermethrin. Israeli Journal of Aquaculture = Bamidgeh. 2002;**54**:99-103. DOI: http://hdl.handle.net/10524/19056

[43] Ahmad F, Ali SS, Shakoori AR. Sublethal effects of Danitol (Fenpropathrin), a synthetic pyrethroid, on freshwater Chinese grass carp, *Ctenopharyngodon idella*. Folia Biologica (Kraków). 1995;**43**:151-159

[44] Galloway T, Handy R. Immunotoxicity of organophosphorous pesticides. Ecotoxicology. 2003;**12**:345-363. DOI: 10.1023/A:1022579416322

[45] Beaman JR, Finch R, Gardner H, Hoffmann F, Rosencrance A, Zelikoff JT. Mammalian immunoassays for predicting the toxicity of malathion in a laboratory fish model. Journal of Toxicology and Environmental Health. 1999;**56**:523-542. DOI: 10.1080/009841099093501755

[46] Hontela A. Interrenal dysfunction in fish from contaminated sites: *In vivo* and *in vitro* assessment. Environmental Toxicology and Chemistry. 1998;**17**: 44-48. DOI: 10.1002/etc.5620170107

[47] Tang Y, Boutilier PG. White muscle intracellular acid-base and lactate status following exhaustive exercise: A comparison between freshwater and sea water adapted rainbow trout. Journal of Experimental Biology. 1991;**156**:153-171

[48] Perry SF, Walsh PJ. Metabolism of isolated fish gill cells: Contribution of epithelial chloride cells. Journal of Experimental Biology. 1989;**144**:507-520

[49] Schulte PM, Mayes CD, Hochachka PW. Intergrating metabolic pathways in post exercise recovery of white muscle. Journal of Experimental Biology. 1992;**166**:181-1854

[50] Weber JM. Pathways for oxidative fuel provision to working muscles. Ecological consequences of maximal supply limitations. Experientia. 1992;**48**:551-564. DOI: 10.1007/BF01920239

[51] Rojik I, Nemso K, Borose L. Morphological and biochemical studies on the liver, kidney, and gill of fishes affected by pesticides. Acta Biologica Hungarica. 1983;**34**(1):81-92

[52] Ganathy V, Reddy SD, Reddy SL, Shankaraiah V. Effect of hexachlorocyclohexane on biochemical composition of the fish *Channa punctatus*. Journal of Ecotoxicology and Environmental Monitoring. 1994;**4**(1):15-20

[53] Bálint T, Szegletes T, Szegletes Z, Halasy K, Nemcsók J. Biochemical and subcellular changes in carp exposed to the organophosphorus methidathion and the pyrethroid deltamethrin. Aquatic Toxicology. 1995;**33**:279-295. DOI: 10.1016/0166-445X(95)00029-4

[54] Jee LH, Masroor F, Kang JC. Responses of cypermethrin-induced stress in haematological parameters of Korean rockfish, *Sebastes schlegeli* (Hilgendorf). Aquaculture Research. 2005;**36**:898-905. DOI: 10.1111/j.1365-2109.2005.01299.x

[55] Datta M, Kaviraj A. Ascorbic acid supplementation of diet for reduction of deltamethrin induced stress in freshwater catfish *Clarias gariepinus*. Chemosphere. 2003;**53**:883-888. DOI: 10.1016/S0045-6535(03)00557-5

[56] Van Leeuwen CJ, Maas-Diepeveen JL, Overbeek HCM. Sublethal effects of tetramethylthiuram disulfide (thiram) in rainbow trout (*Salmo gairdneri*). Aquatic Toxicology. 2002;**9**(1):13-19. DOI: 10.1016/0166-445X(86)90003-2

[57] Kirby MF, Morris S, Hurst M, Kirby SJ, Neall P, Tylor T, et al. The use of cholinesterase activity In flounder (*Platichthys Flesus*) muscle tissue As a

biomarker of neurotoxic contamination in UK estuaries. Marine Pollution Bulletin. 2000;**40**(9):780-791. DOI: 10.1016/S0025-326X(00)00069-2

[58] Carr RL, Strauss DL, Chambers JE. Inhibition and aging of catfish brain acethylcholinesterase following exposure to two phosphorothionate insecticides and their active metabolites. Journal of Toxicology and Environmental Health. 1995;**45**:325-336. DOI: 10.1080/15287399509531999

[59] Costin D, Staicu AS, Dinu D, Huculeci R, Costache A, Dinischiotu A. Biochemical and histological effects of deltamethrin exposure on the gills of *Carassius auratus gibelio* (*Pisces Cyprinidae*), Lucrări ştiinţifice zootehnie şi biotehnologii. Timişoara. 2007;**40**(1):23-28

[60] Clark JM, Matsumura F. Two different types of inhibitory effects of pyrethroids on nerve Ca- and Ca+Mg ATPase in the squid, *Loligo pealei*. Pesticide Biochemical Physiology. 1982;**4**:232-238. DOI: 10.1016/0048-3575(82)90104-3

[61] Li ZH, Grabic R, Li P, Machova J, Velisek J, Randak T. Effects of exposure to sublethal propiconazole on the antioxidant defense system and Na^+-K^+-ATPase activity in brain of rainbow trout, *Oncorhynchus mykiss*. Aquatic Toxicology. 2010;**98**(3):297-303. DOI: 10.1016/j.aquatox.2010.02.017

[62] Cengiz EI. Gill and kidney histopathology in the freshwater fish *Cyprinus carpio* after acute exposure to deltamethrin. Environmental Toxicology and Pharmacology. 2006;**21**:1093-1096. DOI: 10.1016/j.etap.2006.03.006

[63] Cengiz EI, Unlu E. Sublethal effects of commercial deltamethrin on the structure of the gill, liver and gut tissues of mosquitofish, *Gambusia affinis*: A microscopic study. Environmental Toxicology and Pharmacology. 2006;**21**:246-253. DOI: 10.1016/j.etap.2005.08.005

[64] Skidmore JF. Toxicity of zinc compounds to aquatic animals with special reference to fish. The Quarterly Review of Biology. 1964;**39**:227-248. DOI: 10.1086/404229

[65] Burton DT, Jones AH, Cairns J. Acute zinc toxicity to rainbow trout (*Salmo gairneri*): Confirmation of the hypothsis that death is related to tissue hypoxia. Journal of the Fisheries Research Board of Canada. 1972;**29**:1463-1466. DOI: 10.1139/f72-225

[66] Muir DCG, Rawn GP, Townsend BE, Lockhart WL. Bioconcentration of cypermethrin, deltamethrin, fenvalerate and permethrin by *Chironomus tentans* larvae in sediment and water. Environmental Toxicology and Chemistry. 1985;**4**:51-61. DOI: 10.1139/f72-225

[67] Smith TM, Stratton GW. Effects of synthetic pyrethroid insecticides on nontarget organisms. Research Reviews. 1986;**97**:93-119. DOI: 10.1007/978-1-4612-4934-4_4

[68] Mello GC, Donatti L, Rudniki CA, Fanta E. Hepatic alterations in the fish *Rhamdia quelen* contaminated with Folidol 600®. Ecotoxicology and Environmental Safety. 2008;**71**(3):821-829. DOI: 10.1016/j.ecoenv.2008.02.001

[69] Sarkar B, Chatterjee A, Adhikari S, Ayyappan S. Carbofuran- and cypermethrin-induced histopathological alterations in the liver of *Labeo rohita* (Hamilton) and its recovery. Journal of Applied Ichthyology. 2005;**21**:131-135. DOI: 10.1111/j.1439-0426.2004.00590.x

[70] Ali D, Nagpure NS, Kumar S, Kumar R, Kushwaha B, Lakra WS. Assessment of genotoxic and mutagenic effects of chlorpyrifos in freshwater fish *Channa punctatus* (Bloch) using micronucleus assay and alkaline single-cell gel

electrophoresis. Food and Chemical Toxicology. 2009;**47**(3):650-656. DOI: 10.1016/j.fct.2008.12.021

[71] Di Giulio RT, Washburn PC, Wenning RJ, Winston GW, Jewell CS. Biochemical responses in aquatic animals: A review of determinants of oxidative stress. Environmental Toxicology and Chemistry. 1989;**8**: 1103-1123. DOI: 10.1002/etc.5620081203

[72] Bainy ACD, Saito E, Carvalho PSM, Junqueira VBC. Oxidative stress in gill, erythrocytes, liver and kidney of Nile tilapia (*Oreochromis niloticus*) from a polluted site. Aquatic Toxicology. 1996;**34**:151-162. DOI: 10.1016/0166-445X(95)00036-4

[73] Banaee M, Sureda A, Zohiery F, Hagi BN, Garanzini N. Alterations in biochemical parameters of the freshwater fish, *Alburnus mossulensis*, exposed to sub-lethal concentrations of Fenpropathrin. International Journal of Aquatic Biology. 2014;**2**(2):58-68

[74] Sayeed I, Parvez S, Pandey S, Bin-Hafeez B, Haque R, Raisuddin S. Oxidative stress biomarkers of exposure to deltamethrin in freshwater fish, *Channa punctatus Bloch*. Ecotoxicology and Environmental Safety. 2003;**56**:295-301. DOI: 10.1016/S0147-6513(03)00009-5

[75] Abdollahi M, Ranjbou A, Shadnic S, Nikfar S, Rezaice A. Pesticides andoxidative stress: A review. Medical Science Monitor. 2004;**10**:141-143

[76] Ambali S, Akanbi D, Igbokwe N, Shittu M, Kawu M, Ayo J. Evaluation of subchronic chlorpyrifos poisoning on hematological and serum biochemical changes in mice and protective effect of vitamin C. The Journal of Toxicological Sciences. 2007;**32**:111-120. DOI: 10.2131/jts.32.111

[77] Dandapam ZA, Kavima MR, Jayaseelan L, Dommen A. Oxidative damage in intermediate syndrome of acute organophosphorous poisoning. Indian Journal of Medical Research. 2003;**17**:253-259

[78] Fabra A, Angelini J, Donolo A, Permigiain M, Castro S. Biochemical alteration in Bradyrhizobium Sp USDA 3187 induced by the fungicide mancozeb. Antonie van Leeuwenhoek. 2004;**73**(3):223-228. DOI: 10.1023/A:1000987524112

[79] Edwards R, Millburn P, Hutson DH. Comparative toxicity of cys-cypermethrin in rainbow trout, frog, mouse and quail. Toxicology and Applied Pharmacology. 1986;**84**:512-522. DOI: 10.1016/0041-008X(86)90256-5

[80] Levine SL, Oris JT, Denison MS. Modulation of CYP1A expression in rainbow trout by a technical grade formulation of propiconazole. Environmental Toxicology and Chemistry. 1999;**18**:2565-2573. DOI: 10.1002/etc.5620181125

[81] Mauck WL, Olson LE, Marking LL. Toxicity of natural pyrethrins and five pyrethroids to fish. Archives of Environmental Contamination and Toxicology. 1976;**4**:18-29. DOI: 10.1007/BF02221012

[82] Macek KJ, Hutchinson C, Cope OB. The effects of temperature on the susceptibility of bluegills and rainbow trout to selected pesticides. Bulletin Environmental Contamination and Toxicology. 1969;**4**:174-183. DOI: 10.1007/BF01560960

[83] Marinescu AG, Drăghici O, Ponepal C, Păunescu A. The influence of fungicide (Dithane M-45) on some physiological indices in the prussian carp (*Carassius auratus* gibelio Bloch). In: Proceedings 35th IAD Conference; 19-23 April; Novi-Sad, Serbia and Montenegro; Limnological Reports. Vol. 35, 2004. pp. 209-214

[84] Marienescu AlG, Draghici O, Ponepal C, Păunescu A. Research on the changes of some physiological parameters in several fish species under the action of the Thiametoxame insecticide. In: Proceedings of the 36th International Conference IAD; 4-8 September 2006, Klosterneuburg Vienna; 2006. pp. 163-167

[85] Ponepal MC, Păunescu A, Drăghici O, Marinescu AG. Research on the changes of some physiological parameters in several fish species under the action of the thiametoxame insecticide. In: Proceedings 36th International Conference of IAD; 4-8 September; Vienna; 2006. pp. 163-167

[86] Ponepal MC, Păunescu A, Marinescu AlG, Drăghici O. Research on the changes of some physiological parameters in Prussian carp under the action of the Reldan insecticide. In: Proceedings of the Second International Conference "Research People and Actual Tasks on Multidisciplinary Sciences"; 10-12 June; Lozenec, Bulgaria. Vol.1. 2009. pp. 209-212

[87] Ponepal MC, Păunescu A, Drăghici O, Marinescu AG. Research on the changes of some physiological parameters under the action of the Tiradin fungicide in prussian carp and crayfish. Annals. Food Science and Technology. 2009;**10**(1):267-270

[88] Ponepal MC, Păunescu A, Marinescu AG, Drăghici O. The changes of some physiological parameters in Prussian carp under the action of the tilt fungicide. Bulletin USAMV Animal Science and Biotechnologies. 2009;**66**(1-2):47-52

[89] Ponepal MC, Păunescu A, Marinescu AG, Drăghici O. Effect of the Fungicide Chlorothalonil (Bravo) on some physiological parameters in Prussian Carp. Lucrări Ştiinţifice USAMV Iaşi, Seria Horticultură. 2009;**52**:1157-1162. http://www.uaiasi.ro/revista_horti/arhiva.php?an=2009

[90] Ponepal MC, Păunescu A, Marinescu Al G, Drăghici O. Research on the changes of some physiological parameters in several fish species under the action of the Talstar insecticide. Analele Universităţii din Oradea, Fascicula Biologie. Tom. 2010;**XVII**(1):175-179

[91] Ponepal MC, Păunescu A, Drăghici O, Marinescu Al G. Changes of some physiological parameters in Prussian carp under the action of some fungicides. Analele Universităţii Ovidius, Seria Biologie-Ecologie. 2010;**14**:83-88

[92] Ponepal MC. Investigaţii ecofiziologice şi ecotoxicologice la unele specii de peşti din râul Argeş [thesis]. Universitatea din Pitesti. 2011

[93] Schuytema GS, Nebeker AV. Effects of ammonium nitrate, sodium nitrate and urea on red-legged frogs: Pacific treefrogs and African clawed frogs. Bulletin of Environmental Contamination and Toxicology. 1999;**63**:357-364

[94] Pollet I, Bendell-Young LI. Amphibians as indicators of wetland quality in wetlands formed from oil sands effluent. Environmental Toxicology and Chemistry. 2000;**19**:2589-2597. DOI: 10.1002/etc.5620191027

[95] Marco A, Cash D, Belden LK, Blaustein AR. Sensitivity to urea fertilization in three amphibian species. Archives of Environmental Contamination and Toxicology. 2001;**40**:406-409. DOI: 10.1007/s002440010190

[96] Bridges CM. Predator-prey interactions between two amphibian species: Effects of insecticide exposure. Journal Aquatic Ecology. 1999;**33**:205-211. DOI: 10.1023/A:1009922825444

[97] Bridges CM. Long-term effects of pesticide exposure at various life stages of the southern leopard frog (*Rana sphenocephala*). Journal Archives of

Environmental Contamination and Toxicology. 2000;**39**:91-96

[98] Sampath K, Kennedy IJJ, James R. Pesticide impact on excretory physiology of the common frog, *Rana tigrina* (Daud) tadpoles. Bulletin of Environmental Contamination and Toxicology. 2002;**68**:652-659

[99] Broomhall S, Shine R. Effects of the insecticide endosulfan and presence of congeneric tadpoles on australian treefrog (*Litoria freycineti*) tadpoles. Archives of Environmental Contamination and Toxicology. 2003;**45**:221-226

[100] Tomizawa M, Casida J. Selective toxicity of neonicotinoids attributable to specificity of insect and mammalian nicotinic receptors. Annual Review of Entomology. 2002;**48**:339-364. DOI: 10.1146/annurev.ento.48.091801.112731

[101] Tomizawa M, Casida J. Neonicotinoid insecticide toxicology: Mechanisms of selective action. Annual Review of Pharmacology and Toxicology. 2004;**45**:247-268.DOI:10.1146/annurev.pharmtox.45.120403.095930

[102] Gendron AD, Marcogliese DJ, Barbeau S, Christin MS, Brousseau P, Ruby S, et al. Exposure of leopard frogs to a pesticide mixture affects life history characteristics of the lungworm *Rhabdias ranae*. Journal Oecologia. 2003;**135**: 469-476. DOI: 10.1007/s00442-003-1210-y

[103] Greulich K, Pflugmacher S. Uptake and effects on detoxication enzymes of cypermethrin in embryos and tadpoles of amphibians. Journal Archives of Environmental Contamination and Toxicology. 2004;**47**:489-495

[104] Howe CM, BeMGill M, Pauli BD, Helbing CC, WeMGy K, Veldhoen N. Toxicity of glyphosate-based pesticides to four north American frog species. Environmental Toxicology and Chemistry. 2004;**23**(8):1928-1938

[105] Cy C, Hathaway KM, Folt CL. Multiple stress effects of vision herbicide, pH, and food on zooplankton and larval amphibian species from forest wetlands. Environmental Toxicology and Chemistry. 2004;**23**(4):823-831

[106] Relyea RA. Growth and survival of five amphibian species exposed to combinations of pesticides. Environmental Toxicology and Chemistry. 2004;**23**(7):1737-1742

[107] Shaolong F, Zhiming K, Xinming W, Lirong Z, Pingan P. Acute toxicity and genotoxicity of two novel pesticides on amphibian, *Rana N. Hallowell*. Chemosphere. 2004;**56**:457-463. DOI: 10.1016/j.chemosphere.2004.02.010

[108] Relyea RA. The lethal impact of Roundup® on aquatic and terrestrial amphibians. Ecological Applications. 2005;**15**:1118-1124

[109] Relyea RA, Schoeppner NM, Hoverman JT. Pesticides and amphibians: The importance of community context. Ecological Applications. 2005;**15**(4):1125-1134

[110] Hofer R, Lackner R, Lorbeer G. Accumulation of toxicants in tadpoles of the common frog (*Rana temporaria*) in High Mountains. Archives of Environmental Contamination and Toxicology. 2005;**49**:192-199. DOI: 10.1007/s00244-004-0188-8

[111] Seifert J, Stollberg J. Antagonism of a neonicotinoid insecticide imidacloprid at neuromuscular receptors. Environmental Toxicology and Pharmacology. 2005;**20**(1):18-21. DOI: 10.1016/j.etap.2004.09.011

[112] Honda H, Tomizawa M, Casida J. Neonicotinoid metabolic activation and inactivation established with coupled nicotinic receptor-CYP3A4 and -aldehyde oxidase systems. Toxicology Letters. 2006;**161**(2):108-114

[113] Zafeiridou G, Geronikaki A, Papaefthimiou C, Tryfonos M, Kosmidis EK, Theophilidis G. Assessing the effects of the three herbicides acetochlor, 2,4,5-trichlorophenoxyacetic acid (2,4,5-T) and 2,4-dichlorophenoxyacetic acid on the compound action potential of the sciatic nerve of the frog (*Rana ridibunda*). Chemosphere. 2006;**65**:1040-1048. DOI: 10.1016/j.chemosphere.2006.03.032

[114] Casco VH, Izaguirre MF, Marín L, Vergara MN, Lajmanovich RC, Peltzer P, et al. Apoptotic cell death in the central nervous system of Bufo arenarum tadpoles induced by cypermethrin. Cell Biology and Toxicology. 2006;**22**:199-211. DOI: 10.1007/s10565-006-0174-1

[115] Relyea RA. The effects of pesticides, pH, and predatory stress on amphibians under mesocosm conditions. Journal Ecotoxicology. 2006;**15**:503-511. DOI: 10.1007/s10646-006-0086-0

[116] Thompson DG, Wojtaszek BF, Staznik B, Chartrand DT, Stephenson GR. Chemical and bio-monitoring to assess potential acute effects of vision herbicide on native amphibian larvae in forest wetlands. Environmental Toxicology and Chemistry. 2004;**23**:843-849

[117] Edginton AN, Rouleau C, Stephenson GR, Boermans HJ. 2,4-D butoxyethyl ester kinetics in embryos of *Xenopus laevis*: The role of the embryonic jelly coat in reducing chemical absorption. Archives of Environmental Contamination and Toxicology. 2007;**52**:113-120. DOI: 10.1007/s00244-005-0215-4

[118] Vonesh RJ, Buck CJ. Pesticide alters oviposition site selection in gray treefrogs. Oecologia. 2007;**154**:219-226

[119] Gurushankara HP, Krishnamurthy SV, Vasudev V. Effect of malathion on survival, growth, and food consumption of indian cricket frog (*Limnonectus limnocharis*) tadpoles. Archives of Environmental Contamination and Toxicology. 2007;**52**:251-256. DOI: 10.1007/s00244-006-0015-5

[120] Henson-Ramsey H, Kennedy-Stoskopf S, Levine JF, Taylor SK, Shea D, Stoskopf MK. Acute toxicity and tissue distributions of Malathion in *Ambystoma tigrinum*. Archives of Environmental Contamination and Toxicology. 2008;**55**:481-487. DOI: 10.1007/s00244-007-9091-4

[121] Relyea RA, Diecks N. An unforeseen chain of events: Lethal effects of pesticides on frogs at sublethal concentrations. Ecological Applications. 2008;**18**(7):1728-1742

[122] Kang HS, Gye MC, Kim MK. Effects of endosulfan on survival and development of *Bombina orientalis* (Boulenger) embryos. Bulletin of Environmental Contamination and Toxicology. 2008;**81**:262-265. DOI: 10.1007/s00128-008-9475-9

[123] Kang HS, Park CJ, Gye MC. Effects of molinate on survival and development of *Bombina orientalis* (Boulenger) embryos. Bulletin of Environmental Contamination and Toxicology. 2009;**82**:305-309. DOI: 10.1007/s00128-008-9602-7

[124] Relyea RA, Jones D. The toxicity of roundup original MAX to 13 species of larval amphibians. Environmental Toxicology and Chemistry. 2009;**28**(9):2004-2008. DOI: 10.1897/09-021.1

[125] Jones DK, Hammond JI, Relyea RA. Very highly toxic effects of endosulfan across nine species of tadpoles: Lag effects and family level sensitivity. Environmental Toxicology and Chemistry. 2009;**28**(9):1939-1945. DOI: 10.1897/09-033.1

[126] Hayes TB, Collins A, Lee M, Mendoza M, Noriega N, Stuart AA. Hermaphroditic, demasculinized frogs after exposure to

the herbicide atrazine at low ecologically relevant doses. Proceedings of the National Academy of Sciences of the United States of America. 2002;**99**:5476-5480. DOI: 10.1073/pnas.082121499

[127] Hayes T, Haston K, Tsui M, Hoang A, Haeffele C, Vonk A. Atrazine induced hermaphrodism at 0.1 ppb in American leopard frogs (*Rana pipiens*): Laboratory and field evidence. Environmental Health Perspectives. 2003;**111**:568-575. DOI: 10.1289/ehp.5932

[128] Hecker M, Giesy JP, Jones PD, Jooste AM, Carr JA, Solomon KR. Plasma sex steroid concentrations and gonadal aromatase activities in African clawed frogs (*Xenopus laevis*) from South Africa. Environmental Toxicology and Chemistry. 2004;**23**:1996-2007

[129] Coady KK, Murphy MB, Villeneuve DL, Hecker M, Jones PD, Carr JA. Effects of atrazine on metamorphosis, growth, laryngeal and gonadal development, aromatase activity, and sex steroid concentrations in *Xenopus laevis*. Ecotoxicology and Environmental Safety. 2005;**62**:160-173. DOI: 10.1016/j.ecoenv.2004.10.010

[130] Hecker M, Kim WJ, Park JW, Murphy MB, Villeneuve D, Coady KK. Plasma concentrations of estradiol and testosterone, gonadal aromatase activity and ultrastructure of the testis in *Xenopus laevis* exposed to estradiol or atrazine. Aquatic Toxicology. 2005;**72**:383-396. DOI: 10.1016/j.aquatox.2005.01.008

[131] Murphy MB, Hecker M, Coady KK, Tompsett AR, Jones PD, du Preez LH. Atrazine concentrations, gonadal gross morphology and histology in ranid frogs collected in Michigan agricultural areas. Aquatic Toxicology. 2006;**76**:230-245. DOI: 10.1016/j.aquatox.2005.09.010

[132] Storrs SI, Semlitsch RD. Variation in somatic and ovarian development: Predicting susceptibility of amphibians to estrogenic contaminants. General and Comparative Endocrinology. 2008;**156**:524-530. DOI: 10.10166/j.ygcen2008.03.001

[133] Storrs SI, Tillitt DE, Rittenhouse TA, Semlitsch RD. Behavioral response and kinetics of terrestrial atrazine exposure in American toads (*Bufo americanus*). Archives of Environmental Contamination and Toxicology. 2009;**57**:590-597. DOI: 10.1007/s00244-009-9292-0

[134] Williams BK, Semlitsch RD. Larval responses of three midwestern anurans to chronic, low-dose exposures of four herbicides. Archives of Environmental Contamination and Toxicology. 2010;**58**(3):819-827. DOI: 10.1007/s00244-009-9390-z

[135] Budischak AS, Belden KL, Hopkins AW. Relative toxicity of malathion to trematode-infected and noninfected *Rana palustris* tadpoles. Archives of Environmental Contamination and Toxicology. 2009;**56**:123-128. DOI: 10.1007/s00244-008-9167-9

[136] Paunescu A, Soare LC, Fierascu RC, Fierascu I, Ponepal MC. The influence of six pesticides on physiological indices of *Pelophylax Ridibundus* (Pallas, 1771). Bulletin of Environmental Contamination and Toxicology. 2018;**100**(3):376-383. DOI: 10.1007/s00128-018-22777-9

[137] Knutson MG, Richardson WB, Reineke DM, Gray BR, Parmelee JR, Weick SE. Agricultural ponds support amphibian populations. Ecological Applications. 2004;**14**:669-684

[138] Mazerolle MJ. Drainage ditches facilitate frog movements in a hostile landscape. Landscape Ecology. 2005;**20**:579-590. DOI: 10.1007/s10980-004-3977-6

[139] Herzon I, Helenius J. Agricultural drainage ditches, their biological importance and functioning. Biological Conservation. 2008;**141**:1171-1183. DOI: 10.1016/j.biocon.2008.03.005

Chapter 2

Metabolic Impairments Caused by Pesticides in Mammals and Their Interactions with Other Pollutants

Gema Rodríguez-Moro, Ana Arias-Borrego, Sara Ramírez-Acosta, Francisco Navarro-Roldán, Nieves Abril-Díaz, Rut Fernández-Torre, Miguel Angel Bello-López, José Luis Gómez-Ariza and Tamara García-Barrera

Abstract

The biological systems are exposed to a complex environment in which the contaminants can interact in a synergistic/antagonistic fashion and for this reason, the study of "chemical cocktails" is of great interest to fully understand the final biological effect. To evaluate the final biological response of a pollutant, it is essential to have an adequate analytical methodology that allows the correct monitoring of environmental systems in order to establish their quality, and, when appropriate, the application of corrective measures. Undoubtedly, massive methods "the omics" are among the most efficient current tools. To this end, transcriptomics, proteomics, metabolomics and chemical speciation can provide very useful information, mainly when they are combined. However, the combination of proteomics with metabolomics has some drawbacks as the temporal space is different (i.e. metabolomics gives information about what happens right now, but it can be related with numerous post-translational modifications happened previously). In this sense, it seems that the combination of genomics with metabolomics is easier. Thus, when metabolomics data are interpreted in combination with genomic, transcriptomic and proteomic results, in the so-called systems biology approach, a holistic knowledge of the organism/process under investigation can be achieved.

Keywords: pesticides, mammals, metabolomics, metals, speciation, omics

1. Introduction

The evaluation of the biological response in living organisms against environmental pollution requires the use of environmental bioindicators or laboratory models of increasing complexity (mollusks, crustaceans and rodents) [1, 2]. A mammal should be always included to integrate the diverse biological filters present in humans, like the digestive tract, which regulates the passage of contaminants to be later distributed through the bloodstream to other parts of the body [3]. In addition, cell cultures are also important in these types of studies, especially to translate the effects of pollutants to humans. The use of cell cultures also allows performing many experiments without the difficulties of the experimentation with animals [4].

Until now, there has been a great concern, mainly relying on individual environmental pollutants, to study their potential risks. However, the evaluation of health effects of chemicals by considering data obtained just from a single chemical, leads to over or underestimates the joint toxicity. In this sense, studies concerning the combined effects of pollutants better than the toxicity assessment of single chemicals, reflect the realistic impact of environmental exposures [5]. Thus, the experimentation with animal models to evaluate the biological response against environmental pollutants, and their possible translation to the effects in humans, should be carried out designing experiments that mimic as much as possible the environment. To this end, controlled exposure experiments to "chemical cocktails" that integrate several environmental pollutants of different chemical groups can be a good approximation. Although the design of experiments is significantly more complex, the obtained information is of great interest to evaluate the real effects of contaminants in the environment [6].

On the other hand, the joint toxicity of chemical mixtures can be independent one from each other (additive), stronger (synergistic) or weaker (antagonistic), depending of on the sum of effects from individual exposures or not [7].

The biological response can be evaluated using model organisms, but also with free-living species that can also serve as biomarkers of environmental pollution. Although the use of free-living organisms allows evaluating the biological response taking into account all the existing environmental factors and their interactions, and for this reason is closer to the reality, the main pitfalls are related with the difficulty to isolate the metabolic responses connected with a particular pollutant [8]. Otherwise, the study with model organisms is easier and cheaper, but requires a qualified technician to administrate single xenobiotics or their cocktails and the results can be biased by the selected administration route (e.g., subcutaneous injection), model organism used, etc., [10]. Nevertheless, the study of the biological response in free-living animals is of great interest to validate the results obtained in the exposure experiments with model organisms [11, 13].

2. Interactions of metals and chemical species through antagonistic and synergistic mechanisms

As previously commented, biological systems are exposed to a complex environment in which contaminants can interact by means of antagonistic or synergistic mechanisms making mandatory global studies (i.e.,-omics) to evaluate the biological response with an holistic view [14]. In connection to that, selenium has been claimed as the most important element regarding its antagonistic protective action against numerous contaminants (i.e., pesticides and metals). Likewise, it has been stated that cardiovascular damages caused by mercury can be antagonized by selenium [15], but also the neurotoxicity [16] and renal toxicity [17] caused by this element. Moreover, selenium present also protective character against skin cancer induced by arsenic exposure [18], chromosomal aberrations induced by cadmium [19], oxidative stress and lipid peroxidation caused by organophosphorus pesticides [20]. Moreover, the chemical form or specie of selenium is very important since the essentiality or toxicity is directly related to that. Likewise, some of them, especially selenoproteins are peroxidases or oxidoreductases, which protect against oxidative stress [21–24, 26], while inorganic selenium is toxic at high levels [14]. Our research

Mammal	Interaction	Biological effect	References
Interactions of selenium- and arsenic containing species			
Humans	A S_{etotal}/A_{stotal}	DNA hypomethylation/cancer	[39]
Humans		Se reduces the risk of As-related skin lesions and cancer	[16, 40, 41]
Rats	2−A SeO_3/iAs(III)	As prevents lethal liver damage caused by Se	[42]
Rats		As prevents Se-induced cataracts	[43]
Humans		Skin lesion/skin cancer	[44]
Mice		As prevents carcinogenic effect of Se	[45]
Rats		As prevents carcinogenic effect of Se	[46]
Rats		As protects against the toxicity of Se (growth, mortality rate, pathological condition of the liver)	[47–50]
Rats		As induces mucosal glutathione synthesis, explaining its protective effect against Se	[51]
Dogs		As antagonizes Se-induced subnormal growth and restricted food intake	[52]
Cattle		As protects against Se toxicity	[47, 53]
Hogs		As protects against Se toxicity	[48]
Steers		As protects against Se toxicity	[54]
Mice		Se prevents As-induced cytotoxicity	[55]
Mice	2 − 3−A SeO_3/ AsO_4	Se decreases the ratio of organic/inorganic As	[56]
Hamsters		Se decreases As methylation	[14]
Rats	S SeBet/iAs(III)	Coadministration enhances the tumor-suppressive effect of Se	[13]
Interactions of selenium- and mercury-containing species			
Humans	A Setotal/Hgtotal	Se prevents Hg induced cardiovascular diseases	[15]
Humans	A Setotal/$MeHg^+$	Se inhibits Hg-induced neurotoxicity	[57]
Humans		Se inhibits Hg-induced cardiovascular diseases	[58]
Rats		Se may alter the reproductive and developmental toxicity of MeHg+	[59]
Rats	2−/2+ A SeO_3 Hg	Se antagonizes Hg-induced intestinal necrosis	[60]
Rats		Se prevents Hg renotoxicity	[62]
Mice		Se prolongs the half-lives of Hg-exposed animals	[63]
Rats		Se changes the subcellular Hg distribution	[64]
Mice	2−/+ A SeO_4 MeHg	Se protects against Hg-induced neurotoxicity	[65]
Rats	A SeMet/Hg^{2+}	Se inhibits the effects of Hg on organic activities	[66]
Rats	A SeMet/$MeHg^+$	Se prevents Hg-induced porphyrinuria	[67]
Humans	A SeProt/ Hg^0	Se detoxifies Hg	[68]
Mice	A SeProt/$MeHg^+$	Hg affects the activities of selenoenzymes	[69]
Interactions of selenium and sulfur-containing species			
Sheep	A SeMet/sulfur compounds	More Se is incorporated into wool and plasma protein when dietary S is limiting	[70]

Mammal	Interaction	Biological effect	References
Interactions of selenium with species of elements			
Rats	A Se/Sb	Sb has a partially protective effect against Se toxicity	[49, 52, 71]
Rats	A Se/Bi	Bi has a partially protective effect against Se toxicity	[49, 71]
Human cells	A Se/Cd	Se protects against Cd toxicity	[72]
Rats	A Se/Cd	Se prevents Cd-induced oxidative stress	[73]
	A Se/Cd	Se protects against Cd-induced nephrotoxicity and hepatotoxicity	[74]
Mice	A Se/Cd	Se protects against Cd-induced chromosomal aberrations	[19]
Monkeys	A Se/Cd	Se protects enzyme systems	[75]
Rats	S Se/Cd	Se and Cd affect the hepatic gluconeogenic pathway	[76]
	A Se/Cd	Se partially restores Cd-induced oxidative stress and decreased sperm count and motility	[77]
	A Se/Cd	Se antagonizes the Cd-induced inhibition of hepatic drug metabolism	[78]
	A Se/Cd	Se antagonizes Cd-induced testicular damage	[79]
	A Se/Cd	Hepatoprotective effects of Se against Cd	[80]
	A Se/Ge	Ge partially protects against Se toxicity	[49, 71]
	A Se/Ni	Se may antagonize the deleterious effects of Ni during reproduction	[81, 82]
Mice	S Se/Ag	Se protects against Ag-induced lipid peroxidation in the liver	[83, 84]
	S Se/Te/Hg	Hg retention is increased by pre-administration of Te or Se	[85]
Rats	A W/Se	W partially protects against Se toxicity	[49, 71]

Table 1.
Main interaction of selenium species with other elements.

group has been working with *Mus musculus* mice exposed to the pesticide dichloro-diphenyldichloroethylene (DDE) and selenium [27] to study the join effect in the metabolome, as well as several exposure experiments to one or two metals (metalloids) in *Mus musculus* like Se-Cd [2], Hg-Se [28] and others.

Perhaps, the most studied interaction of selenium is with mercury, which was first investigated in 1967 in rats exposed to mercury chloride and selenite [29]. One of the proposed mechanism for this interaction is the formation of Hg-Se complexes, which can result in an increased whole-body retention of Hg after the co-exposure to both elements [29, 30]. However, although this interaction is well known, the mechanisms related to that remain unsolved. It has been stated that inorganic mercury can be incorporated to selenoproteins, peptides and prosthetic groups of selenoenzymes, by reaction of mercury with the selenol group of selenocysteine (SeCys) [31]. The lower pKa of SeCys makes it more reactive that Cys and for this reason, the former reacts by means of -she with Hg with stronger affinity than -SH groups. Mercury can be also incorporated into selenoproteins which play important roles in the maintenance of cellular homeostasis [26, 32]. To this end, Hg^{2-} can react with Se^{2-} (selenides), selenols or hydrogen selenide to form ternary complexes Hg—Se—S together with glutathione that finally bond to selenoprotein P (SelP) [32–34]. The dysfunction of several proteins, like selenoproteins is in the basis of several diseases like carcinogenesis. In this sense, selenoprotein P accounts for the highest content of selenium in serum and can be also present in other selenoenzymes such as thioredoxin reductase (ThxR), glutathione peroxidase (GPx) and selenoprotein P [35, 36].

Previous studies carried out in our research group using *Mus musculus* mice as a model organism exposed to mercury and selenium demonstrate that the levels of selenoprotein P increase in the liver (extracts from hepatic cytosolic extracts) and serum after Hg exposure, and that selenite supplementation increase the effect [28]. In this context, the synthesis of SelP from selenite increases after mercury exposure since this protein serves as a vehicle for Hg detoxification [37]. This fact is also in good agreement with the decrease observed in the selenometabolites found in serum and the correlative increase in liver, where SelP synthesis takes places from selenite to be later transferred to bloodstream [38]. Moreover, our studies demonstrated that the accumulation of SelP in the liver is higher when the diet is supplemented with selenium. On the other hand, the concentration of selenoalbumin increase in liver and decrease in serum after Hg exposure to transport selenium for the synthesis of SelP [38]. Thus, these results demonstrate the interaction of mercury and selenium in the detoxification process induced by the later, the accumulation of selenoproteins in liver and bloodstream and the homeostasis of elements.

Table 1 shows the main selenium interactions with other metals in mammals.

3. Pesticides, metals and their impairments at molecular level

The joint effect induced by pesticides and metal ions increase, in general, the toxicity (synergism) like DNA damage, mortality rate and reduction in the reproduction rate, as well as changes in enzyme activities [86]. The exposure to environmental pollutants such as polychlorinated biphenyls (PCBs), organochlorine pesticides and heavy metals, has been associated to immunotoxic effects in mammals such as alterations of both the innate and adaptive arms of immune systems, which include aspects of cellular and humoral immunity [87].

Metabolic impairments caused by the join exposure of the pesticide DDE and selenium have been studied using a metabolomics approach [27]. In this study, we conclude that about 70 metabolites are altered in the most metabolically active organs, like liver and kidney, but also in brain, and that they are related with the pathways of oxidative stress, degradation of phospholipidic membrane, β-oxidation and energy metabolism, which confirm the potential of combined metabolomic platforms in environmental studies. Moreover, several metabolites present different response (increase or decrease against the control group) in the organs studied indicating a possible traffic between them. This is the case of liver and kidney, which are the most metabolically active organs and present five metabolites altered with the opposite tendency between them, namely: diacylglycerol (DAG) (18:2/18:1) and ornithine and triacylglycerols (TAG) (18:4/18:4/20:4), TAG (16:0/18:1/22:0) and TAG (18:4/18:4/22:6).

To deep insight into the protective effect of selenium against the toxic effect of DDE, we can compare mice supplemented with DDE, DDE + selenium and selenium against the control group. Selenium counteracts the effect of DDE on seven metabolites because they show a different response among the studied groups when they are compared against the control, which is illustrated in **Figure 1**. These metabolites can be used a biomarkers of the antagonistic interaction between selenium and DDE.

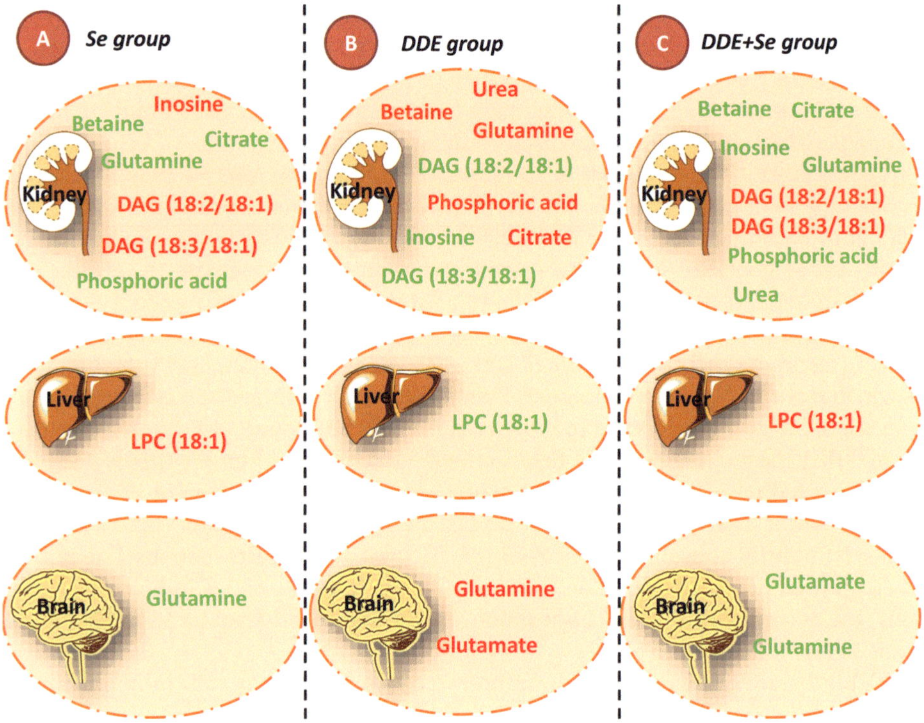

Figure 1.
Different response of metabolites in Mus musculus *mice exposed to: (A) selenium, (B) DDE, (C) selenium and DDE. Red word: Increased; Green word: Decreased.*

It is also remarkable that these antagonistic interactions between DDE and Se usually take place in kidney, since the majority of metabolites that present different response between the mice supplemented with DDE and DDE with Se occur in this organ.

4. Pharmaceutical active compounds, metals and pesticides and their impairments at molecular level

The high consumption of medicines, cosmetic products, as well as pesticides in modern agriculture or plastics in the handling and conservation of food, among others, has led to the appearance in the environment, mainly in soils and aquatic environment, of a series of compounds harmful to the living organisms that inhabit it. These are the well-known "organic micropollutants" (MCOs), a large group of substances that are continuously incorporated into the environment and that, in general, are difficult to eliminate. These substances are classified into two main groups: priority pollutants, and the so-called emerging pollutants (CE). Its detection in the environment has been possible, in many cases, thanks to the development of new and more sensitive analytical technologies that have allowed the detection of these compounds in all types of environmental samples, even in zones, apparently, not subjected to this "contaminant pressure". The analysis of "effluents of wastewater treatment plants" (WWTPs), urban and industrial, has shown, unequivocally, in general, very poor elimination of most of these substances, which is why an incorporation occurs continuously through this way to aquatic and terrestrial ecosystems [88].

Thus, besides metals and pesticides, pharmacologically active compounds (PAC), is a group of emerging contaminants, which are receiving increasing attention because of their potential harmful effects for the environment and human health. The prevention of the emission of priority and emerging pollutants through wastewater treatment plants effluents into the aqueous environment requires the development of new treatment technologies that ensure the quality of receiving water bodies since the actual treatments are deficient and these substances are continuously incorporated into the environment [89]. Actually, analytical methodologies allow determining these substances in almost all samples and at very low levels [88, 90]. All the pollutants present in the aquatic and therefore terrestrial systems, as well as the products that are originated from them by degradation or metabolization, have an inevitable effect on the organisms that inhabit them, it can be highlighted, for example, their presence in coastal sediments or American red crab [90].

The studies related to the effect on biological responses by the presence of chemical cocktails concerning mixtures of pesticides, PACs and metals are scarce [91]. In this context, the effect that the presence of the antibiotic, ciprofloxacin has on the toxicity, distribution and accumulation of copper has been studied in earthworm (*Eisenia fetida*) [92]. However, the study of the biological response of mammals against "chemical cocktails" including metals, pesticides, and PACs has not been performed until now.

5. Omic methodologies to assess health effects of pesticides and other organic micropollutants

It is essential to have an adequate analytical methodology that allows the correct monitoring of environmental systems in order to establish their quality, and, when appropriate, the application of corrective measures. Undoubtedly, massive methods "the omics" are among the most efficient current tools. In this sense, it has been demonstrated that alterations of the homeostatic cycles are shown at the transcriptional level (transcriptomics) [93], by overexpression or inhibition of proteins (proteomics) [94] and by modifications of the metabolic cycles (metabolomics) [1, 95]. On the other hand, it has been stated that approximately 1/3 of proteins need the presence of metals as cofactors to develop their function (metalloproteins) and that metals influence on more than 50% of the proteins [96]. These metals play essential roles due to their catalytic properties or influence the structure of proteins and generally, the genome determines their presence in molecules [97]. Thus, metallomics allows understanding the distribution of elements, concentration at equilibrium of free metallic ions or free elements in a cellular compartment, cell or organism [98] and refers to the identity and/or quantity of metals/metalloids and their species [99]. Likewise, it has been proposed the integration of a global (holistic) view, much more realistic of the processes that takes place in the environment [95].

To this end, genomics decipher the information that determines cell function which is contained in the cellular core, transcriptomics reveals gene expression and proteomics make possible the examination of protein synthesis and cell signaling. On the other hand, in the establishment of transcriptional expression profiles that explain the gene function is critical in **environmental transcriptomics**, and the quantification of gene expression changes at the level of mRNA has proved to be an interesting tool in environmental approaches [100]. The transcriptomes of *M. spretus* mice captured in areas of high industrial and agricultural pollution such as the Domingo Rubio estuary or the industrial pole of Punta del Sebo (Huelva) have been determined, identifying a set of potential biomarkers of environmental contamination [93]. Likewise, the transcriptional profile of contaminants such as the DDE has been determined in controlled exposure studies in the laboratory [101].

On the other hand, proteomics is of great importance to understand cell homeostasis, to perform quantitative analysis and for the identification of potential biomarkers of *in vivo* toxicity. However, the massive number of proteins and post-translational modifications difficult the analysis. Proteomics can also assist genomic and transcriptomic studies in the efficient sequencing of complete genomes and to explain differences in susceptibility induced by polymorphisms. Nowadays is accepted that alterations in gene expression do not always induce adverse health effects due to post-translational modifications, as phosphorylation and glycosylation of proteins that determine their function, environmental factors or multigenic processes (e.g., aging and disease). For this reason, **environmental proteomics** allows understanding the mode of action of pollutants. Likewise, the use of 1st, 2nd and 3rd generation proteomic methods have been used to compare *M. spretus* mice from different areas within the Doñana National Park (PND) and surroundings, the Estero de Domingo Rubio (EDR) and the Estuary of the Guadalquivir, in order to evaluate the effects of the Aznalcóllar accident and the pesticides used around the PND [102, 103]. The information obtained by this omic can be complemented by **oxidative stress and redox proteomics**. The study of reactive oxygen species (ROS) related to oxidative damage caused by contaminants is a valuable tool in environmental studies [94]. Although most amino acids are sensitive to oxidation, the thiol group of Cys affects them especially [95] so its analysis allows determining the redox state of the thiol groups of proteins [104, 105], so its analysis allows determining the redox state of the thiol groups of proteins.

Environmental metallomics and chemical speciation requires its own methodology, generally based in the combined use of inorganic mass spectrometry (inductively coupled plasma mass spectrometry, ICP-MS) and organic MS, using a previous chromatographic separation step in order to preserve the integrity of the metal in terms of its union and location in the molecule [106]. Our research group have extensive experience in this field, especially in its projection to the environmental studies [2], and the biological response, at the metallomic and metabolomic levels of the *Mus musculus* mice exposed to As [99], Cd [2] and Hg [103]. In addition, studies with free living animals in Doñana National Park and surroundings based on these methodologies have also been performed using the *Mus spretus* mice [11] and the crab *Procambarus clarkii*.

In addition, in 1999, J. Nicholson defines metabonomics as "the quantitative measurement of the dynamic multiparametric metabolic response of living systems to pathophysiological stimuli or genetic modification" [104], and metabolomics as the measurement of all the metabolites in a specified biological sample [104]. Likewise, metabonomics allows understanding the variation in low molecular mass molecules, namely metabolites, which are the last action mechanism of the organisms, but the additional mentioned "-omics" sciences are related to cellular macromolecules. Then, the last step in the omics, directly in connection with the phenotype, **environmental metabolomics** allows obtaining a global view of the metabolic fingerprint of the biological systems exposed to contaminants, providing information at the same time that the interactions between the contaminants with the living organisms [105]. Numerous studies have been carried out on different rodents, and several exposure experiments on the *Mus musculus* laboratory mice can be highlighted. As an example, several studies carried to study of the metabolomic response of the mouse.

M. musculus exposed to inorganic As [99], Cd [95], Se-Cd [2], Hg [99], Hg-Se [28], DDE-Se [27], As, Cd, Se, Hg, deltamethrin + acrolein.

One of the most critical aspects in metabolomics is sample treatment, to extract as many metabolites as possible [10]. The analysis of biofluids for metabolomics in simpler than the extraction of metabolites from tissues and allows obtaining global information about the state of the organism, but to obtain specific information on a specific organ the direct analysis is mandatory [106].

Finally, metabolomics allows the simultaneous measurement of hundreds of metabolites *in vitro* cell cultures, *in* vivo tissues and even in non-invasive blood and urine biofluids. However, the current drawback of this omic is the standardization of quenching, metabolite extraction procedures as well as the complexity of data analysis and interpretation. Moreover, the influence of factors in the results is important (e.g., age, gender, diet, stress, housing conditions, health status). To overcome these limitations, combination of omics seems to be the best option. However, the temporal space is different in metabolomics and proteomics (i.e., metabolomics gives information about what happens right now, but it can be related with numerous post-translational modifications happened previously). In this sense, it seems that the combination of genomics with metabolomics is easier. Thus, when metabolomics data are interpreted in combination with genomic, transcriptomic and proteomic results, in the so-called systems biology approach, a holistic view of the organism or biological process under investigation can be attained.

6. Concluding remarks

The evaluation of the real impact of a pollutant, and in particular of pesticides can be performed in the different environmental compartments or preferably in mammals to decipher the biological response. In this case, the study can be carried out in free-living animals or in laboratory mice exposed to different pollutants that should be combined to evaluate the biological response of the "chemical cocktail" since they can interact in a synergistic or antagonistic fashion. On the other hand, it is essential to have an adequate analytical methodology that allows the correct monitoring of environmental systems in order to establish their quality, and, when appropriate, the application of corrective measures. Undoubtedly, massive methods "the omics" are among the most efficient current tools.

Acknowledgements

This work has been supported by the projects CTM2015-67902-C2-1-P from the Spanish Ministry of Economy and Competitiveness and P12-FQM-0442 from the Regional Ministry of Economy, Innovation, Science and Employment (Andalusian Government, Spain). Sara Ramírez-Acosta thanks to Spanish Ministry of Economy and Competitiveness for a PhD scholarship (BES-2016-076364). Finally, authors are grateful to FEDER (European Community) for financial support, grants number UNHU13-1E-1611 and UNHU15-CE- 3140.

Conflict of interest

The authors do not have conflict of interest.

Author details

Gema Rodríguez-Moro[1,2,3], Ana Arias-Borrego[1,2,3], Sara Ramírez-Acosta[1,2,3], Francisco Navarro-Roldán[3,5], Nieves Abril-Díaz[2,4], Rut Fernández-Torre[6], Miguel Angel Bello-López[6], José Luis Gómez-Ariza[1,2,3] and Tamara García-Barrera[1,2,3*]

1 Department of Chemistry, Faculty of Experimental Sciences, University of Huelva, Huelva, Spain

2 International Agrofood Campus of Excellence International ceiA3, University of Huelva, Huelva, Spain

3 Research Center of Natural Resources, Health and the Environment (RENSMA), University of Huelva, Spain

4 Department of Biochemistry and Molecular Biology, University of Córdoba, Córdoba, Spain

5 Department of Environmental Biology and Public Heath, Cellular Biology, Faculty of Experimental Sciences, University of Huelva, Spain

6 Department of Analytical Chemistry, Faculty of Chemistry, University of Seville, Seville, Spain

*Address all correspondence to: tamara@uhu.es

References

[1] Gago-Tinoco A, González-Domínguez R, García-Barrera T, Blasco-Moreno J, Bebianno MJ, Gómez-Ariza J-L. Metabolic signatures associated with environmental pollution by metals in Doñana National Park using *P. clarkii* as bioindicator. Environmental Science and Pollution Research. 2014;**21**: 13315-13323. DOI: 10.1007/s11356-014-2741-y

[2] García-Sevillano MA, García-Barrera T, Gómez-Ariza JL. Application of metallomic and metabolomic approaches in exposure experiments on laboratory mice for environmental metal toxicity assessment. Metallomics. 2014;**6**:237-248. DOI: 10.1039/c3mt00302g

[3] Gómez-Ariza JL, Jahromi EZ, González-Fernández M, García-Barrera T, Gailer J. Liquid chromatography-inductively coupled plasma-based metallomic approaches to probe health-relevant interactions between xenobiotics and mammalian organisms. Metallomics. 2011;**3**:566-577. DOI: 10.1039/c1mt00037c

[4] L'Azou B, Passagne I, Mounicou S, Tréguer-Delapierre M, Puljalté I, Szpunar J, et al. Comparative cytotoxicity of cadmium forms ($CdCl_2$, CdO, CdS micro- and nanoparticles) in renal cells. Toxicology Research. 2014;**3**: 32-41. DOI: 10.1039/c3tx50063b

[5] Schnug L, Leinaas HP, Jensen J. Synergistic sub-lethal effects of a biocide mixture on the springtail Folsomia fimetaria. Environmental Pollution. 2014;**186**:158-164. DOI: 10.1016/J.ENVPOL.2013.12.004

[6] Van den Brink PJ, Tarazona JV, Solomon KR, Knacker T, Van den Brink NW, Brock TCM, et al. The use of terrestrial and aquatic microcosms and mesocosms for the ecological risk assessment of veterinary medicinal products. Environmental Toxicology and Chemistry. 2005;**24, 820**. DOI: 10.1897/04-268R.1

[7] Warne MSJ, Hawker DW. The number of components in a mixture determines whether synergistic and antagonistic or additive toxicity predominate: The funnel hypothesis. Ecotoxicology and Environmental Safety. 1995;**31**:23-28. DOI: 10.1006/EESA.1995.1039

[8] Andrews NC. Metal transporters and disease. Current Opinion in Chemical Biology. 2002;**6**:181-186. DOI: 10.1016/S1367-5931(02)00307-1

[9] García-Barrera T, Rodríguez-Moro G, Callejón-Leblic B, Arias-Borrego A, Gómez-Ariza JL. Mass spectrometry based analytical approaches and pitfalls for toxicometabolomics of arsenic in mammals: A tutorial review. Analytica Chimica Acta. 2018;**1000**:41-66. DOI: 10.1016/j.aca.2017.10.019

[10] García-Sevillano MA, González-Fernández M, Jara-Biedma R, García-Barrera T, López-Barea J, Pueyo C, et al. Biological response of free-living mouse *Mus spretus* from Doñana National Park under environmental stress based on assessment of metal-binding biomolecules by SEC-ICP-MS. Analytical and Bioanalytical Chemistry. 2012;**404**:1967-1981. DOI: 10.1007/s00216-012-6274-2

[11] González-Fernandez M, García-Sevillano MA, Jara-Biedma R, Navarro-Roldán F, García-Barrera T, López-Barea J, et al. Use of metallomics in environmental pollution assessment using mice *mus musculus/mus spretus* as bioindicators. Current Analytical Chemistry. 2013;**9**:229-243. DOI: http://www.scopus.com/inward/record.url?eid=2-s2.0-84875305764&partnerID=tZOtx3Y1

[12] García-Barrera T, Gómez-Ariza JL, González-Fernández M, Moreno F, García-Sevillano MA, Gómez-Jacinto V. Biological responses related to agonistic, antagonistic and synergistic interactions of chemical species. Analytical and Bioanalytical Chemistry. 2012;**403**: 2237-2253. DOI: 10.1007/s00216-012-5776-2

[13] Park K, Mozaffarian D. Omega-3 fatty acids, mercury, and selenium in fish and the risk of cardiovascular diseases. Current Atherosclerosis Reports. 2010;**12**:414-422. DOI: 10.1007/s11883-010-0138-z

[14] Choi AL, Budtz-Jørgensen E, Jørgensen PJ, Steuerwald U, Debes F, Weihe P, et al. Selenium as a potential protective factor against mercury developmental neurotoxicity. Environmental Research. 2008;**107**: 45-52. DOI: 10.1016/J.ENVRES.2007.07.006

[15] Beyrouty P, Chan HM. Co-consumption of selenium and vitamin E altered the reproductive and developmental toxicity of methylmercury in rats. Neurotoxicology and Teratology. 2006;**28**:49-58. DOI: 10.1016/J.NTT.2005.11.002

[16] Kolachi NF, Kazi TG, Wadhwa SK, Afridi HI, Baig JA, Khan S, et al. Evaluation of selenium in biological sample of arsenic exposed female skin lesions and skin cancer patients with related to non-exposed skin cancer patients. The Science of the Total Environment. 2011;**409**:3092-3097. DOI: 10.1016/j.scitotenv.2011.05.008

[17] Mukherjee A, Sharma A, Talukder G. Effect of selenium on cadmium-induced chromosomal aberrations in bone marrow cells of mice. Toxicology Letters. 1988;**41**:23-29. DOI: 10.1016/0378-4274(88)90004-5

[18] Milošević MD, Paunović MG, Matić MM, Ognjanović BI, Saičić ZS. The ameliorating effects of selenium and vitamin C against fenitrothion-induced blood toxicity in Wistar rats. Environmental Toxicology and Pharmacology. 2017;**56**:204-209. DOI: 10.1016/j.etap.2017.09.016

[19] Lai IK, Chai Y, Simmons D, Watson WH, Tan R, Haschek WM, et al. Dietary selenium as a modulator of PCB 126-induced hepatotoxicity in male Sprague-Dawley rats. Toxicological Sciences. 2011;**124**:202-214. DOI: 10.1093/toxsci/kfR215

[20] Raines AM, Sunde RA. Selenium toxicity but not deficient or super-nutritional selenium status vastly alters the transcriptome in rodents. BMC Genomics. 2011;**12**. DOI: 10.1186/1471-2164-12-26

[21] Chen J, Berry MJ. Selenium and selenoproteins in the brain and brain diseases. Journal of Neurochemistry. 2003;**86**:1-12. DOI: 10.1046/j.1471-4159.2003.01854.x

[22] Stadtman TC. Biosynthesis and function of selenocysteine-containing enzymes. The Journal of Biological Chemistry. 1991;**266**:16257-16260

[23] Steinbrenner H, Sies H. Protection against reactive oxygen species by selenoproteins. Biochimica et Biophysica Acta, General Subjects. 2009;**1790**:1478-1485. DOI: 10.1016/j.bbagen.2009.02.014

[24] Rodríguez-Moro G, Abril N, Jara-Biedma R, Gómez-Ariza JL, García-Barrera T. Metabolic impairments caused by "chemical cocktails" in mammals using direct infusion triple quadrupole time of flight and gas chromatography mass spectrometry. Metabolomics (n.d.); 2019

[25] García-Sevillano MA, Rodríguez-Moro G, García-Barrera T, Navarro F, Gómez-Ariza JL. Biological interactions between mercury and selenium in

distribution and detoxification processes in mice under controlled exposure. Effects on selenoprotein. Chemico-Biological Interactions. 2015;**229**:82-90. DOI: 10.1016/j.cbi.2015.02.001

[26] Parziek J, Ostadalova I, Kalouskva J, Babichy A, Benes J. The detoxifying effects of selenium. Interrelation between compounds of selenium and certain metals. In: Mertz W, Cornatzer WE, editors. Newer Trace Element Nutrition. New York, USA: Marcel Dekker; 1971. pp. 85-122

[27] Carvalho CML, Lu J, Zhang X, Arnér ESJ, Holmgren A. Effects of selenite and chelating agents on mammalian thioredoxin reductase inhibited by mercury: Implications for treatment of mercury poisoning. The FASEB Journal. 2011;**25**:370-381. DOI: 10.1096/fj.10-157594

[28] Holben DH, Smith AM. The diverse role of selenium within selenoproteins. Journal of the American Dietetic Association. 1999;**99**:836-843. DOI: 10.1016/S0002-8223(99)00198-4

[29] Falnoga I, Tusek-Znidaric M. Selenium-mercury interactions in man and animals. Biological Trace Element Research. 2007;**119**:212-220. DOI: 10.1007/s12011-007-8009-3

[30] Suzuki KT, Sasakura C, Yoneda S. Binding sites for the (Hg-Se) complex on selenoprotein P. Biochimica et Biophysica Acta-Protein Structure and Molecular Enzymology. 1998;**1429**:102-112. DOI: 10.1016/S0167-4838(98)00221-0

[31] Suzuki KT, Ishiwata K, Ogra Y. Incorporation of selenium into selenoprotein P and extracellular glutathione peroxidase: HPLC-ICPMS data with enriched selenite. The Analyst. 1999;**124**:1749-1753. DOI: 10.1039/a906521k

[32] Matés JM, Segura JA, Alonso FJ, Márquez J. Roles of dioxins and heavy metals in cancer and neurological diseases using ROS-mediated mechanisms. Free Radical Biology & Medicine. 2010;**49**:1328-1341. DOI: 10.1016/j.freeradbiomed.2010.07.028

[33] Matés JM, Segura JA, Alonso FJ, Márquez J. Intracellular redox status and oxidative stress: Implications for cell proliferation, apoptosis, and carcinogenesis. Archives of Toxicology. 2008;**82**:273-299. DOI: 10.1007/s00204-008-0304-z

[34] Chen C, Yu H, Zhao J, Li B, Qu L, Liu S, et al. The roles of serum selenium and selenoproteins on mercury toxicity in environmental and occupational exposure. Environmental Health Perspectives. 2006;**114**:297-301. DOI: 10.1289/ehp.7861

[35] Shiobara Y, Suzuki KT. Binding of selenium (administered as selenite) to albumin after efflux from red blood cells. Journal of Chromatography. B, Biomedical Sciences and Applications. 1998;**710**:49-56. http://www.scopus.com/inward/record.url?eid=2-s2.0-0032511207&partnerID=tZOtx3y1 [Accessed: November 13, 2015]

[36] Richard Pilsner J, Hall MN, Liu X, Ahsan H, Ilievski V, Slavkovich V, et al. Associations of plasma selenium with arsenic and genomic methylation of leukocyte DNA in Bangladesh. Environmental Health Perspectives. 2011;**119**:113-118. DOI: 10.1289/ehp.1001973

[37] Chen Y, Hall M, Graziano JH, Slavkovich V, Van Geen A, Parvez F, et al. A prospective study of blood selenium levels and the risk of arsenic-related premalignant skin lesions. 2007;**16**:207-214. DOI: 10.1158/1055-9965.EPI-06-0581

[38] Spallholz JE, Mallory Boylan L, Rhaman MM. Environmental hypothesis: Is poor dietary selenium intake an underlying factor for

arsenicosis and cancer in Bangladesh and West Bengal, India? The Science of the Total Environment. 2004;**323**:21-32. DOI: 10.1016/J.SCITOTENV.2003.09.034

[39] Verret WJ, Chen Y, Ahmed A, Islam T, Parvez F, Kibriya MG, et al. A randomized, double-blind placebo-controlled trial evaluating the effects of vitamin E and selenium on arsenic-induced skin lesions in Bangladesh. Journal of Occupational and Environmental Medicine. 2005;**47**: 1026-1035. DOI: 10.1097/01.jom.0000183095.45050.97

[40] Moxon AL. The effect of arsenic on the toxicity of seleniferous grains. Science. 1938;**88**:81

[41] Shearer TR, Anderson RS, Britton JL. Influence of selenite and fourteen trace elements on cataractogenesis in the rat. Investigative Ophthalmology and Visual Science. 1983;**24**:417-423

[42] Schrauzer GN, White DA, Mcginness JE, Schneider CJ, Bell LJ. Arsenic and cancer: Effects of joint administration of arsenite and selenite on the genesis of mammary adenocarcinoma in inbred female C_3H/St mice. Bioinorganic Chemistry. 1978;**9**: 245-253. DOI: 10.1016/S0006- 3061(78)80005-2

[43] Ip C, Ganther H. Efficacy of trimethylselenonium versus selenite in cancer chemoprevention and its modulation by arsenite. Carcinogenesis. 1988;**9**:1481-1484. DOI: 10.1093/carcin/9.8.1481

[44] Dubois KP, Moxon AL, Olson OE. Further studies on the effectiveness of arsenic in preventing selenium poisoning one figure. The Journal of Nutrition. 1940;**19**:477-482. DOI: 10.1093/jn/19.5.477

[45] Moxon AL. The influence of arsenic on selenium poisoning in hogs. Proceedings of the South Dakota Academy of Science. 1941;**21**:34-36

[46] Moxon DK, Alvin L. The Journal of Nutrition. 1939:477-482

[47] Palmer IS, Thiex N, Olson OE. Dietary selenium and arsenic effects in rats. Nutrition Reports International. 1983;**27**:249-257

[48] Pisciotto PT, Graziano JH. Induction of mucosal glutathione synthesis by arsenic. Biochimica et Biophysica Acta, General Subjects. 1980;**628**:241-243. DOI: 10.1016/0304-4165(80)90371-2

[49] Rhian M, Moxon AL. Chronic selenium poisoning in dogs and its prevention by arsenic. The Journal of Pharmacology and Experimental Therapeutics. 1943;**78**:249-264 http://jpet.aspetjournals.org/content/78/3/249.abstract

[50] Olson OE. Selenium and the organic arsenicals. Feed Age. 1960;**10**:49-50

[51] Moxon AL, Rhian MA, Anderson HD, Olson OE. Growth of steers on seleniferous range. Journal of Animal Science. 1944;**3**:299-309. DOI: 10.2527/jas1944.33299x

[52] Biswas S, Talukder G, Sharma A. Prevention of cytotoxic effects of arsenic by short-term dietary supplementation with selenium in mice *in vivo*. Mutation Research - Genetic Toxicology and Environmental Mutagenesis. 1999;**441**:155-160. DOI: 10.1016/S1383-5718(99)00028-5

[53] Kenyon EM, Hughes MF, Levander OA. Influence of dietary selenium on the disposition of arsenate in the female $B_6C_3F_1$, mouse. Journal of Toxicology and Environmental Health. 1997;**51**:279-299. DOI: 10.1080/00984109708984027

[54] Holmberg RE, Ferm VH. Interrelationships of selenium,

cadmium, and arsenic in mammalian teratogenesis. Archives of Environmental Health: An International Journal. 1969;**18**:873-877. DOI: 10.1080/00039896.1969.10665508

[55] Ip C, Ganther HE. Activity of methylated forms of selenium in cancer prevention. Cancer Research. 1990;**50**: 1206-1211. http://cancerres.aacrjournals.org/content/50/4/1206.abstract

[56] Yoshizawa K, Rimm EB, Morris JS, Spate VL, Hsieh CC, Spiegelman D, Stampfer NJ, Willett WC. Mercury and the risk of coronary heart disease in men. New England Journal of Medicine.2002;**347**:1755-1760

[57] Parízek J, Ošvtádalová I. The protective effect of small amounts of selenite in sublimate intoxication. Experientia. 1967;**23**:142-143. DOI: 10.1007/BF02135970

[58] An-Sik C, Maines MD, Reynolds WA. Inhibition of the enzymes of glutathione metabolism by mercuric chloride in the rat kidney: Reversal by selenium. Biochemical Pharmacology. 1982;**31**:3093-3100. DOI: 10.1016/0006-2952(82)90085-5

[59] Parízek J, Beneš I, Oštádalová I, Babický A, Beneš J, Lener J. The effect of some inorganic and organic compounds of selenium on the metabolism of cadmium and mercury in the rat. Physiologia Bohemoslovaca. 1969;**18**:95-103

[60] Miura K, Imura N. Changes in ultrastructure and subcellular mercury distribution in rat liver after separate and combined administrations of mercuric mercury and selenite. Journal of Applied Toxicology. 1982;**2**:145-150. DOI: 10.1002/jat.2550020305

[61] Glaser V, Nazari EM, Müller YMR, Feksa L, Wannmacher CMD, Rocha JBT, et al. Effects of inorganic selenium administration in methylmercury-induced neurotoxicity in mouse cerebral cortex. International Journal of Developmental Neuroscience. 2010;**28**: 631-637. DOI: 10.1016/J.IJDEVNEU.2010.07.225

[62] Su L, Wang M, Yin S-T, Wang H-L, Chen L, Sun L-G, et al. The interaction of selenium and mercury in the accumulations and oxidative stress of rat tissues. Ecotoxicology and Environmental Safety. 2008;**70**:483-489. DOI: 10.1016/J.ECOENV.2007.05.018

[63] dos Santos APM, Mateus ML, Carvalho CML, Batoréu MCC. Biomarkers of exposure and effect as indicators of the interference of selenomethionine on methylmercury toxicity. Toxicology Letters. 2007;**169**: 121-128. DOI: 10.1016/J.TOXLET.2006.12.007

[64] Falnoga I, Tusek-Znidaric M. Selenium-mercury interactions in man and animals. Biological Trace Element Research. 2007;**119**:212-220. DOI: 10.1007/s12011- 007-8009-3

[65] Watanabe C, Yin K, Kasanuma Y, Satoh H. In utero exposure to methylmercury and Se deficiency converge on the neurobehavioral outcome in mice. Neurotoxicology and Teratology. 1999;**21**:83-88. DOI: 10.1016/S0892-0362(98)00036-1

[66] White CL. Sulfur-selenium studies in sheep. II. Effect of a dietary sulfur deficiency on selenium and sulfur metabolism in sheep fed varying levels of selenomethionine. Australian Journal of Biological Sciences. 1980;**33**:699-708. DOI: 10.1071/BI9800699

[67] Moxon AL, Jensen C. The influence of germanium gallium antimony and some organic arsenicals on the toxicity of selenium. Proceedings of the South Dakota Academy of Science. 1947:21-26

[68] Frisk P, Yaqob A, Lindh U. Indications of selenium protection

against cadmium toxicity in cultured K-562 cells. The Science of the Total Environment. 2002;**296**:189-197. DOI: 10.1016/S0048-9697(02)00080-3

[69] Messaoudi I, El Heni J, Hammouda F, Saïd K, Kerkeni A. Protective effects of selenium, zinc, or their combination on cadmium-induced oxidative stress in rat kidney. Biological Trace Element Research. 2009;**130**:152-161. DOI: 10.1007/s12011-009-8324-y

[70] Flora SJS, Behari JR, Ashquin M, Tandon SK. Time-dependent protective effect of selenium against cadmium-induced nephrotoxicity and hepatotoxicity. Chemico-Biological Interactions. 1982;**42**:345-351. DOI: 10.1016/0009-2797(82)90078-3

[71] Sidhu M, Sharma M, Bhatia M, Awasthi YC, Nath R. Effect of chronic cadmium exposure on glutathione S-transferase and glutathione peroxidase activities in Rhesus monkey: The role of selenium. Toxicology. 1993;**83**:203-213. DOI: 10.1016/0300-483X(93)90102-X

[72] Bell RR, Soliman MRI, Early JL. Acute effects of cadmium and selenium on glucose output from rat liver hepatocytes using various gluconeogenic precursors. Toxicology. 1990;**65**:161-168. DOI: 10.1016/0300-483X(90)90086-V

[73] Messaoudi I, Banni M, Saïd L, Saïd K, Kerkeni A. Involvement of selenoprotein P and GPx4 gene expression in cadmium-induced testicular pathophysiology in rat. Chemico-Biological Interactions. 2010;**188**:94-101. DOI: 10.1016/J.CBI.2010.07.012

[74] Early JL, Schnell RC. Selenium antagonism of cadmium-induced inhibition of hepatic drug metabolism in the male rat. Toxicology and Applied Pharmacology. 1981;**58**:57-66. DOI: 10.1016/0041-008X(81)90115-0

[75] Prohaska JR, Mowafy M, Ganther HE. Interactions between cadmium, selenium and glutathione peroxidase in rat testis. Chemico-Biological Interactions. 1977;**18**:253-265. DOI: 10.1016/0009-2797(77)90012-6

[76] Newairy AA, El-Sharaky AS, Badreldeen MM, Eweda SM, Sheweita SA. The hepatoprotective effects of selenium against cadmium toxicity in rats. Toxicology. 2007;**242**:23-30. DOI: 10.1016/J.TOX.2007.09.001

[77] Käkelä R, Käkelä A, Hyvärinen H. Effects of nickel chloride on reproduction of the rat and possible antagonistic role of selenium. Comparative Biochemistry and Physiology - Part C: Toxicology & Pharmacology. 1999;**123**:27-37. DOI: 10.1016/S0742-8413(99)00006-7

[78] Khandelwal S, Flora SJS, Tandon SK. Nickel-selenium interaction-time dependent biochemical alterations and metal decorporation in rats. Chemico-Biological Interactions. 1990;**75**:341-347. DOI: 10.1016/0009-2797(90)90076-Y

[79] Serafín Muñoz AH, Wrobel K, Gutierrez Corona JF, Wrobel K. The protective effect of selenium inorganic forms against cadmium and silver toxicity in mycelia of *Pleurotus ostreatus*. Mycological Research. 2007;**111**: 626-632. DOI: 10.1016/J.MYCRES.2007.03.002

[80] Rungby J. Silver-induced lipid peroxidation in mice: Interactions with selenium and nickel.Toxicology.1987; **45**:135-142

[81] Khayat A, Dencker L. Interactions between tellurium and mercury in murine lung and other organs after metallic mercury inhalation: A comparison with selenium. Chemico-Biological Interactions. 1984;**50**:123-133. DOI: 10.1016/0009-2797(84)90089-9

[82] Uwizeyimana H, Wang M, Chen W, Khan K. The eco-toxic effects of pesticide and heavy metal mixtures towards earthworms in soil. Environmental Toxicology and Pharmacology. 2017;**55**:20-29. DOI: 10.1016/j.etap.2017.08.001

[83] Desforges J-PW, Sonne C, Levin M, Siebert U, De Guise S, Dietz R. Immunotoxic effects of environmental pollutants in marine mammals. Environment International. 2016;**86**: 126-139. DOI: 10.1016/j. envint.2015.10.007

[84] Camacho-Muñoz D, Martín J, Santos JL, Aparicio I, Alonso E. Occurrence, temporal evolution and risk assessment of pharmaceutically active compounds in Doñana Park (Spain). Journal of Hazardous Materials. 2010; **183**:602-608. DOI: 10.1016/j. jhazmat.2010.07.067

[85] Deblonde T, Cossu-Leguille C, Hartemann P. Emerging pollutants in wastewater: A review of the literature. International Journal of Hygiene and Environmental Health. 2011;**214**: 442-448. DOI: 10.1016/j. ijheh.2011.08.002

[86] Kazakova J, Fernández-Torres R, Ramos-Payán M, Bello-López M. Multiresidue determination of 21 pharmaceuticals in crayfish (*Procambarus clarkii*) using enzymatic microwave-assisted liquid extraction and ultrahigh-performance liquid chromatography-triple quadrupole mass spectrometry analysis. Journal of Pharmaceutical and Biomedical Analysis. 2018;**160**:144-151. DOI: 10.1016/j.jpba.2018.07.057

[87] Cedergreen N. Quantifying synergy: A systematic review of mixture toxicit studies within environmenta toxicology. PLoSONE.2014;9:e96580.DOI:10.1371/ journal.pone.0096580

[88] Huang R, Wen B, Pei Z, Shan X-Q, Zhang S, Williams PN. Accumulation, subcellular distribution and toxicity of copper in earthworm (*Eisenia fetida*) in the presence of ciprofloxacin. Environmental Science & Technology. 2009;**43**:3688-3693. DOI: 10.1021/ es900061t

[89] Abril N, Ruiz-Laguna J, García-Sevillano MA, Mata AM, Gómez-Ariza JL, Pueyo C. Heterologous microarray analysis of transcriptome alterations in Mus spretus mice living in an industrial settlement. Environmental Science & Technology.2014;**48**:2183-2192. DOI: 10.1021/es4053973

[90] Montes-Nieto R, Fuentes-Almagro CA, Bonilla-Valverde D, Prieto-Álamo M-J, Jurado J, Carrascal M, et al. Proteomics in free-living *Mus spretus* to monitor terrestrial ecosystems. Proteomics. 2007;**7**:4376-4387. DOI: 10.1002/pmic.200700409

[91] García-Sevillano MA, Abril N, Fernández-Cisnal R, García-Barrera T, Pueyo C, López-Barea J, et al. Functional genomics and metabolomics reveal the toxicological effects of cadmium in *Mus musculus* mice. Metabolomics. 2015. DOI: 10.1007/ s11306-015-0801-z

[92] Mounicou S, Szpunar J, Lobinski R. Metallomics: The concept and methodology. Chemical Society Reviews. 2009;**38**:1119-1138. DOI: 10.1039/b713633c

[93] Tainer JA, Roberts VA, Getzoff ED. Metal-binding sites in proteins. Current Opinion in Biotechnology. 1991;**2**:582-591. DOI: 10.1016/0958- 1669(91)90084-I

[94] Williams RJ. Chemical selection of elements by cells. Coordination Chemistry Reviews. 2001;**216–217**: 583-595. DOI: 10.1016/S0010-8545(00) 00398-2

[95] García-Sevillano MA, García-Barrera T, Navarro F, Gómez-Ariza JL. Analysis of the biological response of mouse liver (*Mus musculus*) exposed to As_2O_3 based on integrated -omics approaches. Metallomics. 2013;**5**: 1644-1655. DOI: 10.1039/c3mt00186e

[96] Abril N, Ruiz-Laguna J, Osuna-Jiménez I, Vioque-Fernández A, Fernández-Cisnal R, Chicano-Gálvez E, et al. Omic approaches in environmental issues. Journal of Toxicology and Environmental Health. Part A. 2011;**74**: 1001-1019. DOI: 10.1080/15287394.2011.582259

[97] Morales-Prieto N, Ruiz-Laguna J, Abril N. Dietary Se supplementation partially restores the REDOX proteomic map of *M. spretus* liver exposed to p,p′-DDE. Food and Chemical Toxicology. 2018;**114**:292-301. DOI: 10.1016/j.fct.2018.02.047

[98] Bonilla-Valverde D. Contaminación de doñana: biomarcadores bioquímicos y proteomica en el ramon moruno (*Mus spretus*) y en el gorrion comun (*Passer domesticus*). Universidad de Córdoba. 2006

[99] Abril N, Chicano-Gálvez E, Michán C, Pueyo C, López-Barea J. ITRAQ analysis of hepatic proteins in free-living *Mus spretus* mice to assess the contamination status of areas surrounding Doñana National Park (SW Spain). The Science of the Total Environment. 2015;**523**:16-27. DOI: 10.1016/j.scitotenv.2015.03.116

[100] Morales-Prieto N, Abril N. REDOX proteomics reveals energy metabolism alterations in the liver of *M. spretus* mice exposed to p,p′-DDE. Chemosphere. 2017;**186**:848-863. DOI: 10.1016/j.chemosphere.2017.08.057

[101] Fernández-Cisnal R, Alhama J, Abril N, Pueyo C, López-Barea J. Redox proteomics as biomarker for assessing the biological effects of contaminants in crayfish from Doñana National Park. The Science of the Total Environment. 2014;**490**:121-133. DOI: 10.1016/j.scitotenv.2014.04.117

[102] Sperling M, Karst U. Metallomics: An emerging interdisciplinary science. Analytical and Bioanalytical Chemistry. 2013;**405**:1789-1790. DOI: 10.1007/s00216-012-6619-x

[103] García-Sevillano MA, García-Barrera T, Navarro F, Gailer J, Gómez-Ariza JL. Use of elemental and molecular-mass spectrometry to assess the toxicological effects of inorganic mercury in the mouse *Mus musculus*. Analytical and Bioanalytical Chemistry. 2014;**406**:5853-5865. DOI: 10.1007/s00216-014-8010- 6

[104] Nicholson JK, Lindon JC, Holmes E. "Metabonomics": Understanding the metabolic responses of living systems to pathophysiological stimuli *via* multivariate statistical analysis of biological NMR spectroscopic data. Xenobiotica. 1999;**29**:1181-1189. Available from: http://www.scopus.com/inward/record.url?eid=2-s2.0-0032694577&partnerID=tZOtx3Y1

[105] García-Sevillano MA, García-Barrera T, Gómez-Ariza JL. Development of a new column switching method for simultaneous speciation of selenometabolites and selenoproteins in human serum. Journal of Chromatography A. 2013;**1318**: 171-179. DOI: 10.1016/j.chroma.2013.10.012

[106] Masson P, Alves AC, Ebbels TMD, Nicholson JK, Want EJ. Optimization and evaluation of metabolite extraction protocols for untargeted metabolic profiling of liver samples by UPLC-MS. Analytical Chemistry;**82**(2010): 7779-7786. DOI: 10.1021/ac101722e

Chapter 3

Environmental Risk of Groundwater Pollution by Pesticide Leaching through the Soil Profile

Gabriel Pérez-Lucas, Nuria Vela, Abderrazak El Aatik and Simón Navarro

Abstract

Adsorption, degradation, and movement are the key processes conditioning the behavior and fate of pesticides in the soil. Six processes that can move pesticides are leaching, diffusion, volatilization, erosion and run-off, assimilation by microorganisms, and plant uptake. Leaching is the vertical downward displacement of pesticides through the soil profile and the unsaturated zone, and finally to groundwater, which is vulnerable to pollution. Pesticides are frequently leached through the soil by the effect of rain or irrigation water. Pesticide leaching is highest for weakly sorbing and/or persistent compounds, climates with high precipitation and low temperatures, and soils with low organic matter and sandy texture. On the contrary, for pesticides with a low persistence that disappear quickly, the risk of groundwater pollution considerably decreases. Different and varied factors such as physical-chemical properties of the pesticide, a permeability of the soil, texture and organic matter content of the soil, volatilization, crop-root uptake, and method and dose of pesticide application are responsible for the leaching rate of the pesticides. Soils that are high in clays and organic matter will slow the movement of water, attach easily to many pesticides, and generally have a higher diversity and population of soil organisms that can metabolize the pesticides.

Keywords: aqueous/soil environment, groundwater vulnerability, pesticide leaching, soil pollution

1. Introduction

Agriculture plays an important socioeconomic role in the European Union (EU). The total agricultural area of the EU-28 was 184.6 million hectares in 2015, which supposes 43.5% of its total land area with France and Spain being the countries with greater cultivated land [1]. Therefore, to protect agricultural production and quality, the use of pesticides is widespread.

Pesticides have important benefits in crop protection, food and material preservation, and disease control although unfortunately can pose undesirable effects on human health and environmental ecosystems. The use of pesticides in agriculture is necessary to combat a variety of pests and diseases that could destroy crops and to improve the quality of the food produced. The main estimated losses in crop yields are due to insect pests (14%), plant pathogens (13% loss), and weeds (13%) [1]. Therefore, pesticides are necessary for agricultural production. Among the different classes of pesticides, the highest percentages of an application are corresponding to herbicides (49%), followed by fungicides and bactericides (27%), and insecticides (19%) [2].

A pesticide also called plant protection product (PPP) is any "substance intended for preventing, destroying, repelling, or mitigating any pest in crops either before or after harvest to prevent deterioration during storage or transport." A more detailed definition can be found in the document by FAO [3]. The term includes compounds such as antimicrobials, defoliants, disinfectants, fungicides, herbicides, insecticides, insect growth regulators, molluscicides, and other minority groups. Pesticide products include both active ingredients and inert ingredients. Active ingredients are used to control pests, diseases, and weeds, while inert ingredients (stabilizers, dyes, etc.) are important for product performance and usability.

Regulation (EC) No 1107/2009 [4] is the legislation concerning the placing of PPPs on the market in the European Union. EFSA's Pesticides Unit is responsible for the EU of risk assessments of active substances used in PPPs, in close cooperation with all EU Member States. The risk assessment of active substances evaluates whether, when used correctly, these substances are likely to have any direct or indirect harmful effects on human or animal health, groundwater quality, and nontarget organisms.

Since the 1940s, synthetic pesticides have been widely applied worldwide to protect agricultural crops from pests and diseases, and their use was increased progressively as increased human population and crop production especially from the *Green Revolution*. During 2016, the worldwide consumption of pesticides reached 4.1 millions of tons of active ingredients, which 51.3% was consumed in Asia, 33.3% in Americas (Northern, Central, and South), 11.8% in Europe, 2.2% in Africa, and 1.4% in Oceania. This consumption originated a pesticide trade higher than 60 billion of US $. **Figure 1** shows the evolution in the use of pesticides during the period 1990–2016 in the world, Europe, the United States of America, and the least developed countries [2].

As can be observed, the consumption was ascending (increasing use) in the worldwide and least developed countries and descending (reduced use) in the most developed areas like EU and USA.

However, many of the pesticides used are chemical compounds that persist in the environment being able to be bioaccumulated through the food web and transported to long distances [5] adversely affecting human health and environment around the world, especially organochlorine pesticides [6]. Toxicity of the compound, amount applied and formulation type, method and time of application and, especially, its mobility and persistence are the main factors involved on the risk when a pesticide is incorporated in the environment. In addition, many of them have been identified as endocrine disruptors (EDs), compounds that alter function (s) of the endocrine system and consequently cause adverse health effects in an intact organism, or its progeny, or subpopulations [7–10]. Humans and wildlife depend on the ability to reproduce and develop normally, which is not possible without a healthy endocrine system. Since the beginning of this century, numerous laboratory studies have added to our understanding of the impact of EDs on human and wildlife health [11, 12] and confirmed the scientific complexity of this issue.

The pollution of soil and water bodies by pesticides used in agriculture can pose an important threat to aquatic ecosystems and drinking water resources. Pesticides can enter in water bodies via point sources or diffuse. Surface waters generally

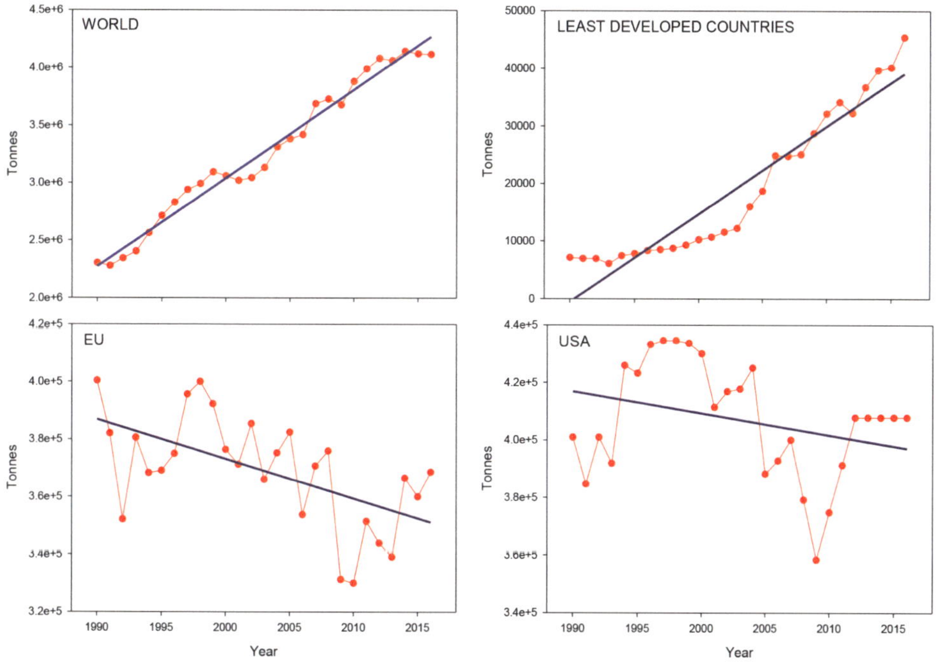

Figure 1.
Evolution of pesticide consumption from 1990 to 2016 (Data obtained from FAOSTAT [2]).

contain a much greater diversity of compounds compared to groundwater although this may be simply a function of the limited amount of groundwater monitoring rather than a surface occurrence. However, according to Directive 2006/118/EC [13], groundwater is the largest body of fresh water in the EU. Concretely, Europe confronts serious episodes of groundwater pollution with agriculture being the biggest polluter. About 60% of European citizens rely on groundwater for drinking water purposes, and its use is threatened by the leaching of pesticides and nitrates due to agricultural practices. In addition, groundwater is used for drinking water by more than 50% of the people in the USA, including almost everyone who lives in rural areas.

Infiltration through riverbeds and riverbanks and leaching through the soil and unsaturated zone are the main diffuse pesticide input paths into groundwater [14, 15]. Therefore, groundwater resources are vulnerable to pollution [16]. Although no universally accepted definition has been contributed for groundwater vulnerability, the National Research Council of USA [17] defines it as "the likelihood for contaminants to reach a specified position in the groundwater system after introduction at some location above the uppermost aquifer." In this context, pesticide residues have been detected in groundwater bodies in the EU [18] and USA [19] at higher levels in some cases than the drinking water limit established by the EU (0.1 mg L^{-1} for individual pesticide and 0.5 mg L^{-1} for $\sum$ pesticides). In this way, the Directive 2009/128/EC [20] was named to protect human health and the environment from possible risks associated with the use of pesticides. The aim of this directive is to achieve a sustainable use of pesticides in the EU by reducing the risks and impacts of pesticide use on human health and the environment and promoting the use of Integrated Pest Management (IPM) and alternatives, such as nonchemical techniques. When pesticides are used, appropriate risk management measures should be established and low-risk pesticides, as well as biological control measures, should be considered in the first place. According to FAO, integrated pest management (IPM) is "an ecosystem approach to crop production and protection that combines different management strategies and practices to grow healthy crops and minimize the use of pesticides" [21]. Other definitions of IPM according to the US EPA [22] involve "an effective and environmentally sensitive approach to pest management that relies on a combination of common-sense practices." IPM, therefore, utilizes the best mix of control tactics for a given pest problem such as host resistance, chemical, biological, cultural, mechanical, sanitary, and mechanical controls using each technique a different set of mechanisms for suppressing populations [23].

2. Soil: fundamental concepts related to pesticide leaching

Defining soil is always a hard task due to its high heterogeneity, the complex processes involved, and quite often its own use. The soil taxonomy defines the soil as *a natural body comprised of solids (minerals and organic matter), liquid, and gases that occurs on the land surface, occupies space, and is characterized by one or both of the following: horizons, or layers, that are distinguishable from the initial material as a result of additions, losses, transfers, and transformations of energy and matter or the ability to support rooted plants in a natural environment* [24]. Soil structure refers to units composed of primary particles. Seven structural classes are recognized in soils: platy, prismatic, columnar, blocky, granular, wedge, and lenticular.

2.1 Soil profile

A soil profile is a vertical section of a soil, showing horizons (layers running parallel to the surface) and parent material. **Figure 2** shows a drawing of a vertical section of soil.

Soil horizons differ in different easily seen soil properties (color, texture, structure, and thickness) and other less visible (chemical and mineral content, consistency, and reaction). The *O horizon* is the layer containing organic materials such as surface organisms, twigs, and dead leaves. It has different levels of decomposition (minimal, moderately, highly, and completely decomposed organic matter). This horizon is often black or dark brown in color, because of its organic content. The roots of small grass are found in this layer. The *A horizon* (also known as the root zone) constitutes the topsoil. It is typically made of sand, silt, and clay with high levels of organic matter and is highly vulnerable to erosion by wind and water. The *B horizon* contains high concentrations of clay, iron, aluminum, and carbonates. Other specific subhorizons will be mentioned, as needed. For example, a B horizon may have several parts if their characteristics such as texture or color change with depth (denoted as Bt1, Bt2, Btg). The *C horizon* is mainly made up of broken bedrock without organic material. It contains geologic material and cemented sediment and there is little activity. The *R horizon* is bedrock (granite, basalt, and limestone), a compacted and cemented material due to the weight of the overlying horizons.

2.2 Soil composition

Although an infinite variety of substances may be found in soil, four basic components constitute it: minerals (45%), organic matter (5%), air (25%), and water (25%). The voids in the soil are known as pore space, and there are two kinds of pores: matrix and nonmatrix pores. Matrix pores are typically smaller than nonmatrix pores in fine- and medium-textured soils.

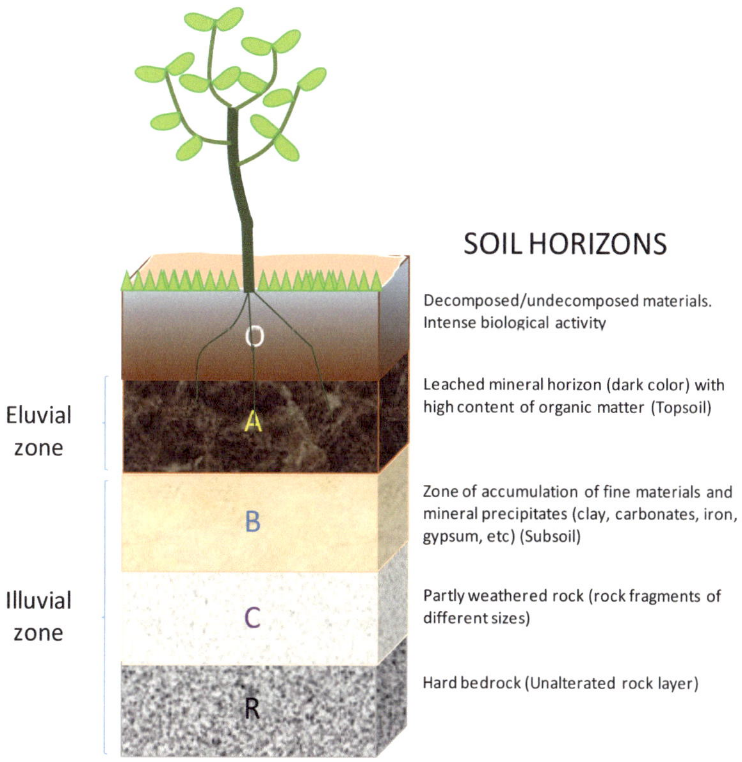

Figure 2.
Schematic drawing of the soil profile.

Air and water are in the pores contained between the solid particles of the soil. The pore sizes vary from very fine (<1 mm) to very coarse (≥10 mm). The ratio of air-/water-filled pore space vary seasonally, weekly, and even daily, depending on water additions through precipitation, flow, groundwater discharge, and flooding. According to suction and gravimetric water contents as defined by suction, three water state classes can be defined: dry (>1500 kPa), moist (≤1500 to >1.0 kPa), and wet (≤1.0 kPa). Natural drainage class refers to the frequency and duration of wet periods. Different drainage classes include excessively drained, somewhat excessively drained, well-drained, moderately well-drained, somewhat poorly drained, poorly drained, very poorly drained, and subaqueous [24].

The mineral portion of soil is divided into a fine fraction (<2 mm in diameter) and larger soil particles (>2 mm in diameter) known as rock fragments. Three particle-size classes integrate the fine fraction: sand (2–0.05 mm), silt (0.05–0.002 mm), and clay (<0.002 mm). These particles differ in their effects on soil drainage and their relative capacity to available hold water for uptake by plants. Texture can be defined as the relative combination of sand, silt, and clay in a soil. Thereby, 12 soil textural classes are represented on the USDA soil texture triangle as can be seen in **Figure 3** [24].

On the other hand, soil organic matter (SOM) is a complex mixture of different substances containing fresh deposits of plants and organisms and humus, a fraction of stable organic compounds mainly humic and fulvic acids that are resistant to further rapid decomposition. An important physical property of SOM is its ability to absorb large quantities of water. The mass and volume of water that can be absorbed by SOM often exceed the mass and volume of the SOM itself. In addition, SOM has a much higher CEC than clays and can also form complexes with metals and organic materials like pesticides, sometimes rendering them immobile [25, 26].

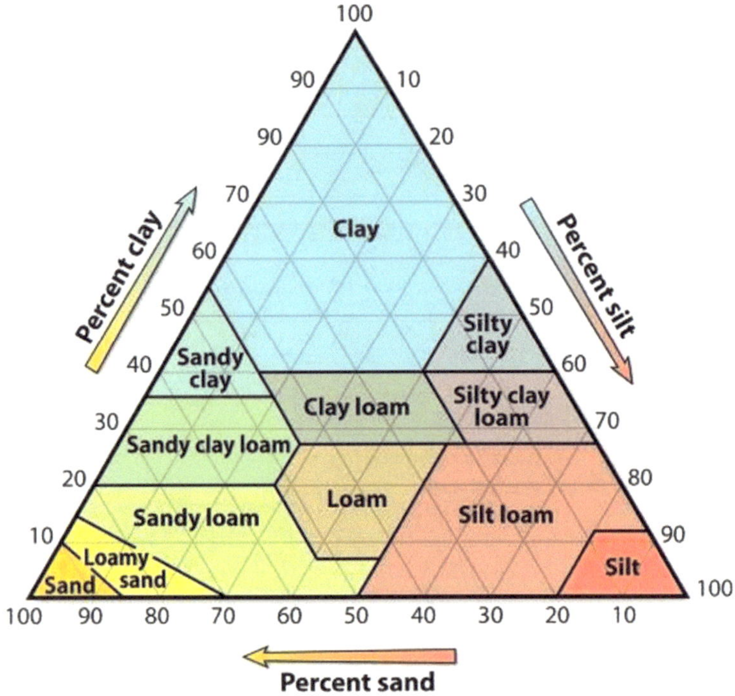

Figure 3.
Possible textural classes of the soil.

3. Behavior and fate of pesticide residues in the soil

In addition to accidental or intentional discharges, the presences of pesticides in agricultural soils mainly have two origins: (i) treatments applied to the aerial part of crops to combat pests, when approximately 50% of the product (insecticides and fungicides, and some herbicides) may reach the soil and (ii) the soil itself is directly treated (insecticides, nematicides, disinfectants, and mainly herbicides), which will obviously lead to a higher concentration in the same [27]. To understand the behavior of a pesticide, it is essential to have the appropriate analytical tools capable of determining residual concentrations in different media (plant, soil, and water) and the main metabolites that can appear. Analytical procedures typically involve a number of equally relevant steps for *sampling, sample preparation, isolation of the target compounds, identification, and quantification* mainly by gas (GC-MS) and liquid chromatography (LC-MS) coupled to mass spectrometry and other minority techniques such as capillary electrophoresis (CE), immunochemical methods (ICMs), electrochemical methods (EMs), chemiluminescence (CL) or ion mobility spectrometry (IMS), and *data processing* [28, 29].

The fate of pesticides in the soil depends on many processes responsible for their mobility and persistence [30, 31]. Persistence may be defined as *the tendency of a pesticide to conserve its molecular integrity and chemical, physical, and functional characteristics for a certain time after being released into the soil*. The half-life time (**$t_{1/2}$**) is the term commonly used to assess persistence (i.e., the time required for a pesticide to degrade to one-half of its initial amount in the soil). The typical half-life to consider a pesticide as persistent is more than 100 days, while nonpersistent pesticides have less than 30 days. Therefore, moderately persistent pesticides have **$t_{1/2}$** ranged from 30 to 100 days [32]. From an environmental point of view, persistent pesticides are undesirable because some of them are intrinsically toxic and deleteriously affect human, domesticated animals, agricultural crops, wildlife, fish and other aquatic organisms, or microorganisms. Some recalcitrant (i.e., nonbiodegradable) pesticides are not toxic at the levels found in the soil, but they can reach hazardous levels due to biomagnifications through the natural food chains. For this reason, it is very important to know the process by which a pesticide is degraded in order to determine whether it will accumulate in the soil or pass into groundwater and whether it will persist in either.

Once a pesticide is applied to soil, it will most likely follow one of three pathways: (i) adhering to soil particles (mainly organic matter and clays), (ii) degrading by organisms and/or free enzymes, and (iii) moving through the soil with water. From the physical-chemical data of adsorption, mobility, and degradation obtained in the laboratory, it is possible to predict with a high degree of reliability the behavior of pesticides in the soil. For this, different guidelines have been proposed by OECD to study adsorption [33], degradation [34], and leaching [35]. **Figure 4** shows the schematic behavior of pesticides in the soil.

Adsorption that may be chemical (electrostatic interactions) or physical (van der Waals forces) is the result of the electrical attraction between charged particles, pesticide molecules (sorbate), and soil particles (adsorbent). Pesticide molecules that are positively charged are attracted to negatively charged particles on clays and organic matter. Chemical reactions between unaltered pesticides or their metabolites often lead to the formation of strong bonds (chemisorption) resulting in an increase in the persistence of the residues in the soil, while causing it to lose its chemical identity.

Degradation generally happens gradually through the formation of one or more metabolites and takes place through photochemical, chemical, and/or microbiological processes. Photodegradation refers to the decomposition induced by radiant energy (ultraviolet/visible light range) on pollutants and is only relevant at the soil

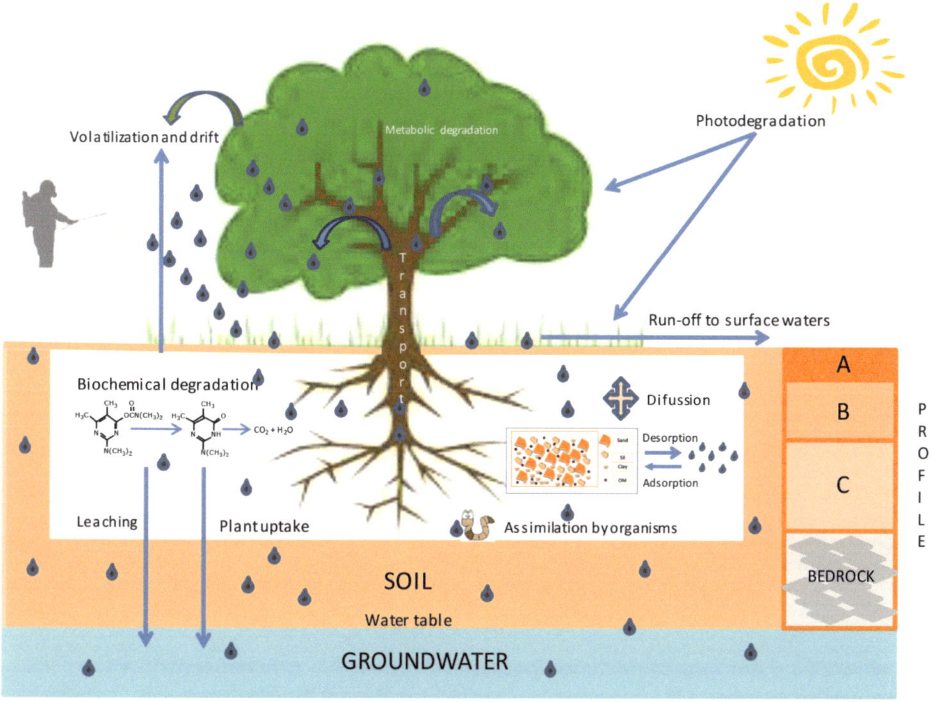

Figure 4.
Behavior and fate of pesticides in the soil.

surface. The solar light may be absorbed by the pollutant, resulting in the formation of by-products, or does not have a direct effect on the pollutant but acts on other substances (photosensitizers) that will promote the degradation of pesticides [36]. Chemical (hydrolysis, oxidation, aromatic hydroxylation, etc.) and biological processes are closely linked and it is difficult to distinguish between them. For this, the process is commonly called biochemical degradation.

The transformations that pesticides may suffer in the soil are many and varied. Besides the characteristics of the pesticide, other factors such as the colloidal composition, texture and moisture content of the soil, the number of microorganisms present (including bacteria and fungi), etc., play a key role. *Biodegradation* can be defined as *a process by which microbial organisms transform or alter, through metabolic or enzymatic action, the structure of pesticides present in the soil* [37]. The metabolic pathways from natural metabolic cycles have enabled the microorganisms to degrade pesticides in the soil although many of them are recalcitrant pesticides. Whereas biodegradable pesticides are broken down within days or weeks by soil microorganisms, recalcitrant pesticides remain for long periods (years or even decades) in the soil.

By a total degradation of a pesticide (mineralization), CO_2, salts, and water are formed, and parts of the chemical are built into new molecular structures in the soil humus or in biomass (bound residues). The terms *free* and *bound* residues were coined to indicate that the former can be readily extracted from soil without altering their chemical structures, whereas the latter are resistant to such extraction [38]. However, the distinction between these two fractions is not always clear, because while they are in the soil, even the extractable residues are not entirely free from any form of binding because they may be adsorbed to the soil solid phases and, therefore, show reduced bioavailability and degradation. According to Roberts [39], bound residues are *chemical species originating from pesticides, used according to good agricultural practice, that is, unextracted by methods which do not significantly change the chemical nature of these residues*. Twelve years later, according to IUPAC, Fuhr et al. [40] proposed a modification to the existing definition of bound residues: *Compounds in soils, plant or animals, which persist in the matrix in the form of the parent substance or its metabolite(s) after extraction*.

Knowledge of the kinetics of biochemical degradation is essential to the evaluation of the persistence of pesticides. Pesticide degradation was described using simple first-order (SFO) kinetics for much time, and it is still the most common mathematical description of pesticide degradation in the scientific literature. However, in some cases, this model is not appropriate. The FOCUS (FOrum for the Coordination of pesticide fate models and their USe) degradation kinetic expert group, supported by the European Commission, came up with two alternative equations for pesticide degradation in soil. Both are based on first-order kinetics although composed of several processes [41]. The alternative equations are the First Order Multi-Compartment (FOMC) equation and the Double First Order in Parallel (DFOP) equation.

$$C_t = C_0 e^{-kt} \text{ (SFO)} \tag{1}$$

$$C_t = C_0 \left(1 + \frac{t}{\beta}\right)^{-\alpha} \text{ (FOMC)} \tag{2}$$

$$C_t = C_1 e^{-k_1 t} + C_2 e^{-k_2 t} \text{ (DFOP)} \tag{3}$$

where C_t = amount of pesticide present at time t, k = rate constant for the degradation process, C_0 = amount of pesticide at time 0 (initial amount), β = parameter determined by the variation in k values, α = positional parameter, C_1 = amount of pesticide at time 0 in the first compartment, k_1 = rate constant for degradation in the first compartment, C_2 = amount of pesticide at time 0 in the second compartment, and k_2 = rate constant for degradation in the second compartment.

Finally, pesticide transfer refers to the movement of pesticides from their site of application. Five processes that can move pesticides are diffusion, volatilization, leaching, erosion and runoff, assimilation by microorganisms, and absorption by plants. Diffusion can be verified in the gaseous and liquid phases, or in the air of the inter-solid phase. The pesticide is transferred through the soil from one zone where it is more concentrated to another where it is less. The volatilization of pesticides from the soil and their subsequent dispersion in the atmosphere is a common occurrence and is perhaps the most important route by which pesticides dissipate. Once volatilized, a pesticide can move in air currents away from the treated surface, a phenomenon known as vapor drift. The soil can be act as a conveyor of the pesticide when its particle is moved from one place to another through the effects of wind or runoff, leading in certain cases to the contamination of surface waters (rivers, seas, and lakes). Runoff determines the movement of water over a sloping surface that occurs when water is applied faster than it enters the soil. Pesticides carried by surface runoff from agricultural areas are a significant portion of the pesticide pollutant loading rates to surface water bodies. Absorption of pesticides by a target and nontarget organisms (bioaccumulation) is quite variable and it is influenced by species characteristics, environmental conditions, and by the chemical-physical properties of both the pesticide and the soil. Pesticide uptake by plants depends on the environmental conditions and the physical-chemical properties of the soil and pesticides and it is influenced by plant species, growth stage, and intended use. Leaching is the vertical downward displacement of pesticides through the soil profile and the unsaturated zone, and finally to groundwater. Pesticide leaching is highest for weakly sorbing and/or persistent compounds, climates with high precipitation and low temperatures (which leads to high groundwater recharge) and sandy-soils with low organic matter.

Figure 5 summarizes major factors (pesticide and soil properties, site conditions, and management practices) affecting the fate of pesticides in the soil [32].

3.1 Leaching process

Nowadays, the study of pesticide leaching represents an important field of research concerning environmental pollution. A large number of papers published

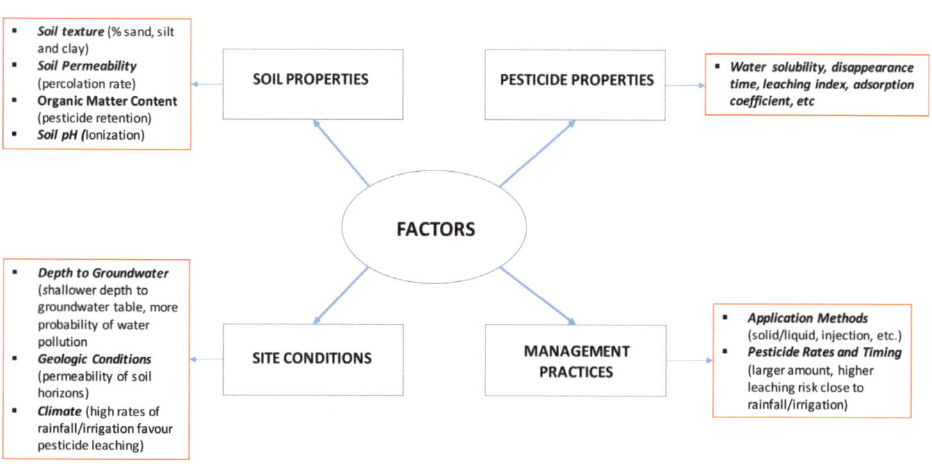

Figure 5.
Factors affecting the fate of pesticides in the soil.

from the beginning of this century to the current moment confirm this interest. A review to the literature extracted from The Web of Science™ (www.isiknowledge.com) managed by Thomson Reuters (Philadelphia, USA) using the keywords *pesticides* AND *leaching* AND *soil* shows about 2500 papers in the period considered.

Leaching constitutes an environmental risk because they can reach the water table and contaminate shallow groundwater and deeper aquifers. However, for pesticides with a low persistence that disappear quickly, the risk of groundwater pollution considerably decreases.

Two different types of flow are associated with pesticide leaching: (i) preferential flow, related to water that flows rapidly through large voids, root channels, and cracks and (ii) matrix flow, due to the slow movement of pesticide/water through the small pores of the soil having in this case more time to contact soil particles [42].

Pesticides are frequently leached through the soil by the effect of rain or irrigation water but for this to happen, the product must be sufficiently soluble in water. The pesticide may be displaced, dissolved, suspended, or simply emulsified in water. Water movement concerns rates of flow into and within the soil and the related amount of water that runs off and does not enter the soil. Infiltration is the process of downward water entry into the soil. Three infiltration stages may be differentiated: (i) steady ponded, (ii) preponded, and (iii) transient ponded. Water that is moving at a high velocity can better carry pesticides of high molecular weight and has the potential to move them farther.

3.1.1 Influential factors

The factors (chemical, physical, and biological) influencing the leaching rate of the pesticides are varied including among others, physical-chemical properties of the pesticide, permeability of the soil, texture and organic matter content of the soil, volatilization, crop-root uptake, and method/dose of pesticide application. Also important is climate change. Pesticide leaching can be affected directly by climate change due to variations in temperature and precipitation patterns or indirectly by any change in the agroecosystem caused by changes in land use, modified application timings, or the use of different pesticides against new invasive pests, diseases, or weeds [43]. Regarding direct effects, increased temperatures should in principle increase pesticide degradation rates, which will, in turn, reduce the risk of leaching although also increase desorption (endothermic process) favoring the liberation of pesticides from soil colloids. On the other hand, an increase in rainfall leads to an increased risk of pesticide leaching.

Different soil adsorption models have been developed for different pesticide classes in order to identify the properties governing retention class-specific quantitative structure-property relationship [44]. **Table 1** summarizes the main physical-chemical properties of a pesticide that can affect its leaching rates and the suggested thresholds according to PPDB [45].

The relation between the concentrations of the compound in the solid and liquid phases is known as the distribution coefficient and is directly proportional to the solubility of the pesticide in water and inversely proportional to the organic matter (OM) and clay content of the soil.

$$K_d = \frac{C_a}{C_d} \quad (4)$$

where K_d = coefficient of partition between soil and water (V/M); C_a = amount of pesticide adsorbed per unit of adsorbent mass (M/M); and C_d = concentration of pesticide dissolved (M/V).

Parameter	Thresholds
WS (mg L^{-1})	<50 = low; 50–500 = moderate; >500 = high
Log K_{OW}	<2.7 = low bioaccumulation; 2.7–3 = moderate; >3.0 = high
DT_{50SD} (days)	<30 = nonpersistent; 30–100 = moderately persistent; 100–365 = persistent; >365 = very persistent
DT_{50AP} (days)	<1 = fast; 1–14 = moderately fast; 14–30 = slow; >30 = stable
DT_{50AH} (days)	<30 = nonpersistent; 30–100 = moderately persistent; 100–365 = persistent; >365 = very persistent
GUS index	>2.8 - high leachability; 2.8–1.8 - transition state; <1.8 - low leachability
VP (mPa)	<5 = low volatility; 5–10 = moderately volatile; >10 = highly volatile
H (Pa m^3 mol^{-1})	>100 = volatile; 0.1–100 = moderately volatile; <0.1 = nonvolatile
Log K_{OC}	<1.2 = very mobile; 1.2–1.9 = mobile; 1.9–2.7 = moderately mobile; 2.7–3.6 = slightly mobile; >3.6 = nonmobile
pK_a	pH < pKa neutral state; pH > pKa negative charge

WS: water solubility, K_{OW}: octanol-water partition coefficient; DT: disappearance time; SD: soil degradation; AP: aqueous photolysis; AH: aqueous hydrolysis; GUS: groundwater ubiquity score index; VP: vapor pressure; H: Henry's law constant; K_{OC}: organic carbon normalized sorption coefficient; K_a: acid dissociation constant.

Table 1.
Main physical-chemical properties influencing the leaching of pesticides.

Karickhoff et al. [46] demonstrated the existence of a linear correlation between the coefficient of partition and the soil's organic carbon content:

$$K_{OC} = \left(\frac{K_d}{OC}\right) \times 100 \tag{5}$$

where K_{oc} = soil organic partition coefficient and OC is the organic carbon content (%).

For polar molecules and soils with low OM content and high clay content, Hermosín and Cornejo [47] found a similar correlation:

$$K_{OC} = \left(\frac{K_d}{CC}\right) \times 100 \tag{6}$$

where K_{cc} = clay content partition coefficient and CC = clay content (%).

Both K_{oc} and K_{cc} are linearly correlated with the coefficient of partition between octanol and water (K_{ow}), which indicates the affinity degree of the pesticide for water (low value) or for soil (high value).

Sorption and degradation processes, both influenced by chemical-physical properties of the soils and compounds involved, and weather conditions, mainly affect the movement of water and dissolved pesticides through the soil. According to some authors, adsorption and desorption are the processes that regulate the magnitude and speed of leaching, and a pesticide should not be affected by other processes while it is adsorbed to the humic-argillic complex [48]. The use of clay barriers modified with cationic surfactants has been demonstrated as an effective method to increase the retention of pesticides in soil [49, 50]. The content of organic carbon (OC) is considered as the single largest factor having maximum influence on pesticide degradation, adsorption, and mobility in soil [51]. Therefore, the soil organic adsorption coefficient (K_{OC}) is generally used as a measure of the relative potential mobility of pesticides in soils to describe the partitioning of them in the water/soil/air compartments.

Thus, a possible mitigation measure to reduce pesticide leaching through the soil could be the increase of the OM content of the soil by agronomic practices like the incorporation of crop residues or animal manures to increase sorption of nonionic pesticides [52]. Another option to reduce leaching by matrix flow would be the use of compounds with high/fast sorption. Addition of OC in the form of crop residues, manure, or sludge is a common soil management practice followed in some areas of the Mediterranean Basin. In this zone, high temperature and evapotranspiration, adverse climatic conditions, and soil degradation are responsible for the decrease in plant growth and consequent lack of organic compounds that would improve the soil nutrient status since its addition contributes to enhancement of active humified components (humic and fulvic acids) [53]. Soils of low OC content have a low capacity to avoid pesticide mobility because humic substances are the primary adsorbent materials for pesticides. Nowadays, the addition of organic amendment (OA) to soils is being intensely studied to know its effect on pesticide sorption and its movement through the soil profile in order to minimize the risk of water pollution associated with rapid runoff and leaching. Soil amended with sludge, urban waste compost, composted straw, fly ash, olive oil mill wastes, spent mushroom substrates, or wood residues has been shown to increase pesticide [54–64]. In addition, recent studies have demonstrated the ability of biochar to decrease pesticide leaching to groundwater. The concept to use biochar as a soil amendment is recent but it really comes from the study of very ancient soils in the Basin of Amazon. Biochar can be defined as a carbon-rich solid material produced by heating biomass in an oxygen-limited environment [65]. Biochar is distinguished from charcoal by its use as a soil amendment. Many and varied properties are attributed to biochar such as C sequestration, reduction of N_2O emissions from soil, bioenergy generation, stimulation of soil microorganisms, sorption of pesticides and nutrients, improvement of soil structure and retention of water, and control of soil-borne pathogens [66–68].

The main benefit concerning the sorption of pesticides to OM is that it generally decreases leaching, where it is due to the presence of additional OM in the amended soil but also to the structural changes in the porosity induced by the presence of new OC content [69]. As a part of the OA added, dissolved organic matter (DOM) is incorporated to the soil, which affects movement and sorption of pesticides [70, 71]. Pesticide leaching may be enhanced by pesticide-DOM interactions and competition for sorption sites between pesticides/DOM molecules [72]. Polarity and molecular weight of the pesticides are key factors on the extent and nature of this behavior [73]. Moreover, the microbiological activity is increased by addition of OA to soil, which enhances the biodegradation of pesticides in polluted soils. Therefore, pesticide behavior in amended soil has reported different results because diverse effects have been pointed [74].

3.1.2 Methodology for leaching studies

In addition to thin-layer chromatography (TLC) and reverse-phase LC, other methods are commonly used to assess the potential leaching of pesticides through the soil. These methods include soil columns, outdoor lysimeters, and field studies [75].

The use of the packed column is a valuable tool to analyze pesticide displacement through the soil. OECD [35] and USEPA [76] have standardized methods to study the leaching process. These studies are generally carried out using disturbed soil columns filled with sieved soil (<2 mm). The use of disturbed soil columns has the advantage of obtaining more reproducible results than other methods. Pesticides are applied on the top of the column followed by percolation with a pesticide-free solution after 24–48 h with distilled or deionized water, and preferably with electrolytes as 0.01 M $CaCl_2$ to minimize colloidal dispersion [77].

Outdoor lysimeters were developed to avoid or at least decrease the differences obtained between laboratory and field conditions [78]. In addition, lysimeters having a large surface area can be used to plant crops to assess pesticide behavior under simulated natural conditions and being the water easily collected from the bottom of them. An additional advantage of the lysimeters over laboratory columns is that the seasonal effect of an application on leaching can be evaluated. For contrast, outdoor lysimeter studies may require many replicates to obtain accurate results on pesticide transport due to the variability of profiles. At field scale, groundwater monitoring and terrestrial field dissipation studies can be considered methods that are more realistic to assess the potential risk of the leaching process.

3.1.3 Indexes for pesticide leaching

Many authors have proposed various indices to predict the mobility of pesticides in the soil. The *Koc* value, obtained by using the batch equilibrium method, is simple and one of the most useful indexes for nonionic pesticides for which the leaching potential is indicated by a mobility classification of immobile to very mobile. They are simple index-based screening tools, which use both the physical-chemical properties of pesticides and soil to make a quick evaluation of pesticide leaching potential considering setting threshold values. **Table 2** summarizes some of the main indices published during the last four decades.

Index/reference	Parameters/equation/interpretation criteria
Hamaker's RF [79]	$R_F = \frac{1}{\{1+(K_{OC} * f_{OC} * \rho_b * (\theta^{-0.67} - 1)\}}$ *$R_F = 0.64$–1: high; $R_F = 0.35$–0.64: moderate; $R_F = 0.1$–0.35: low; $R_F < 0.1$: very low*
McCall's [80]	K_{OC} *0–50, very high; 50–150, high; 150–500, medium; 500–2000, low; 2000–5000, slight; > 5000, immobile*
Briggs's RF [81]	$\mathrm{Log}(1/R_F - 1) = Log(K_{OW}) + Log(OM) - 1.33$ *$R_F = 0.90$–1.0: class 5 (very high); $R_F = 0.89$–0.65: class 4 (high); $R_F = 0.64$–0.35: class 3 (moderate); $R_F = 0.34$–0.10: class 2 (low); $R_F = 0$–0.09: class 1 (very low)*
LEACH [82]	$LEACH = \frac{S_w * t_{½}}{V_p * K_{OC}}$ *Comparison (lower values; lower leaching potential)*
Cohen's [83]	Soil $DT_{50} > 2$–3 weeks; hydrolysis $DT_{50} > 25$ weeks; aqueous photolysis $DT_{50} > 1$ week; soil $Koc < 300$; soil $K_d < 5$; $K_H < 10^{-2}$; WS > 30 mg L^{-1} or field leaching at >75–90 cm *High leaching potential*
Hornsby index [84]	$HI = \left(\frac{K_{OC}}{t_{½}}\right) * 10$ *$HI \leq 10$: high; $HI \geq 2000$: low*
AF [85]	$RF = \left[1 + \frac{\rho_b * f_{OC} * K_{OC}}{\theta_{FC}} + \frac{\theta_g * K_H}{\theta_{FC}}\right]$ $AF = \exp\left[\frac{-0.693 * d * RF * \theta_{FC}}{q * t_{½}}\right]$ *$AF = 0$ to -1: high; $AF = -1$ to -2: moderate; $AF = -2$ to -3: low; $AF = -3$ to -4: very low; $AF < -4$: nonleachable*
GUS [86]	$GUS = [4 - \log(K_{OC})] * \log(t_{½})$ *$GUS > 2.8$: leachable; $GUS = 1.8$–2.8: intermediate; $GUS < 1.8$: nonleachable*

Index/ reference	Parameters/equation/interpretation criteria
LPI [87]	$RF = \left[1 + \frac{\rho_b * f_{OC} * K_{OC}}{\theta_{FC}} + \frac{\theta_g * K_H}{\theta_{FC}}\right] LPI = \frac{1000 * t_{½} * q}{0.693 * RF * Z}$ *LPI > 90: very high; LPI = 75–89: high; LPI = 50–74: moderate; LPI = 25–49: low; LPI = 0–24: very low*
SNV [88]	1. Water solubility greater than 3 ppm; 2. Organic carbon normalized soil sorption coefficient (K_{OC}) < 1900 mL g^{-1}; 3. Hydrolysis DT_{50} > 14 days; 4. Aerobic soil metabolism DT_{50} > 610 days; 5. Anaerobic soil metabolism DT_{50} > 9 days. *If the Boolean expression "(1 or 2) and (3 or 4 or 5)" is TRUE = potential leacher*
PLP index [89]	$PLP_{value} = \frac{R * F * t_{½}}{K_{OC}} PLP_{index} = (\log PLP_{value})(14.3) + 57$ *PLP_{index} = 90–100: very high; PLP_{index} = 70–89: high; PLP_{index} = 50–69: moderate; PLP_{index} = 30–49: low; PLP_{index} = 0–29: very low*
GWCP [90]	$GWCP = \frac{PLP + SLP}{2}$ *GWCP > 150: high; GWCP = 75–150: moderate; GWCP < 75: low*
AFT & AFR [91]	$AFT = \ln AF/(-0.693) AFR = \ln AFT + k$ *Comparison (lower values; lower leaching potential)*
LIX [92]	$LIX = \exp(-k * K_{OC}) or\ LIX = \exp\left(-\frac{0.693}{t_{½}} * K_{OC}\right)$ *LIX = 1: high leachable; LIX = 0.1–1: leachable; LIX = 0–0.1: transition; LIX = 0: nonleachable*
LIN [93]	$LIN = -0.531 \log K_{OW} + 0.518 \log S_w - 0.495 \log K_{OC} - 0.023 \log V_p - 0.452 \log K_H$ *Comparison (lower values; lower leaching potential)*
M. LEACH [94]	$M.LEACH = \frac{S_w * t_{½}}{K_{OC}}$ *Comparison (lower values; lower leaching potential)*
GLI [94]	$GLI = 0.579\ LIN + 0.558\ GUS + 0.595\ M.\ LEACH$ GLI > 1: high; GLI = −0.5 to 1: medium; GLI < −0.5: low
DTK [95]	$DTK = e^{-depth}(a \ln t_{½} - b \ln K_{OC})$ *Leaching value > 0 = potential leacher; leaching value 0 = nonleacher*
VI [16]	$VI = \frac{200 * k * \theta_{FC}}{d * \rho_b * (\%OM)} * \left(\frac{t_{½}}{K_{OC}}\right) * F_{DGW}$ *Comparison (lower values; lower leaching potential)*

$t_{1/2}$: half-life (days); Θ: volumetric soil water content; Z or d: depth to groundwater (m); q: net groundwater recharge rate (m/day); ρ_b: soil bulk density (kg/m^3); OM: organic matter; S_w: water solubility (mg/L); V_P: vapor pressure (mm Hg); Θ_{FC}: volumetric water content at field capacity; F: fraction of pesticide reaching the soil during application; K_H: Henry constant; K_{ow}: octanol/water partition coefficient; K_{oc}: organic carbon normalized soil sorption coefficient (mL/g organic carbon); Θ_g: gas content; RF: retardation factor; V: volatility (bar); f_{oc}: organic carbon fraction; R: rate of pesticide application (kg/ha).

Table 2.
Main indices published during the last four decades about the risk potential of pesticide leaching for groundwater pollution.

Thus, a pesticide screening can be estimated with relatively few input data need, and therefore, these index-based methods are easy to apply, unlike other models that require very intensive field-based data that are very difficult to obtain in many cases as summarized in the next section.

4. Vulnerability risk of groundwater to pesticide pollution

Groundwater can be defined as the *water located beneath the earth's surface in soil pore spaces and in the fractures of rock formations* [96]. The alteration of the chemical equilibrium established between groundwater and the surface through which it circulates, reflected in the appearance of foreign substances or compounds to which they constitute natural quality, serves as an indicator of human activity. When this alteration constitutes a negative impact on the water bodies or affects the potential of the resource for its subsequent use, it can be called pollution. Unlike what happens in surface waters, the detection of pollution and the evaluation of its effects on groundwater resources present serious difficulties. In the groundwater, degradation of quality is often noticed when the polluting process has affected large areas of the aquifer. The adoption of corrective measures, which are expensive and not always effective, is difficult due to the evolution of the contaminant in the medium and the consequent difficulty in establishing a diagnosis of the cause-effect relationships. Therefore, the vulnerability of an aquifer to pollution indicates the sensitivity of groundwater to an alteration in its quality caused by human activities. In addition to the influence exerted by the unsaturated zone, the vulnerability of groundwater as a consequence of a pollution episode is also conditioned by climatological factors (rainfall and temperature), and others related to the polluting load such as method and place of penetration, mobility, and persistence of the pesticide (**Figure 6**).

In 1996, the EPA developed SCI-GROW (Screening Concentration in Groundwater) as a screening-level tool to estimate drinking water exposure concentrations in groundwater resulting from pesticide use [97]. As a screening tool, SCI-GROW provides conservative estimates of pesticides in groundwater, but it does not have the capability to consider variability in leaching potential of different soils, weather (including rainfall), cumulative yearly applications, or depth to aquifer. In 2004, concurrently, the EPA Office of Pesticide Programs (OPP) and Canada's Pesticide Management Regulatory Authority (PMRA) initiated a project to develop a harmonized approach to modeling pesticide concentrations in groundwater called the Pesticide Root Zone Model (PRZM). After this project was completed, the two agencies recommended PRZM-GW as the harmonized tool for assessing pesticide concentrations in groundwater, which was implemented as an exposure model in 2012 [98].

In addition to these models, there are three traditional methods for assessing groundwater vulnerability to pollution with pesticides and other pollutants: (i) process-based, involving numerical modeling, (ii) statistical, involving correlating

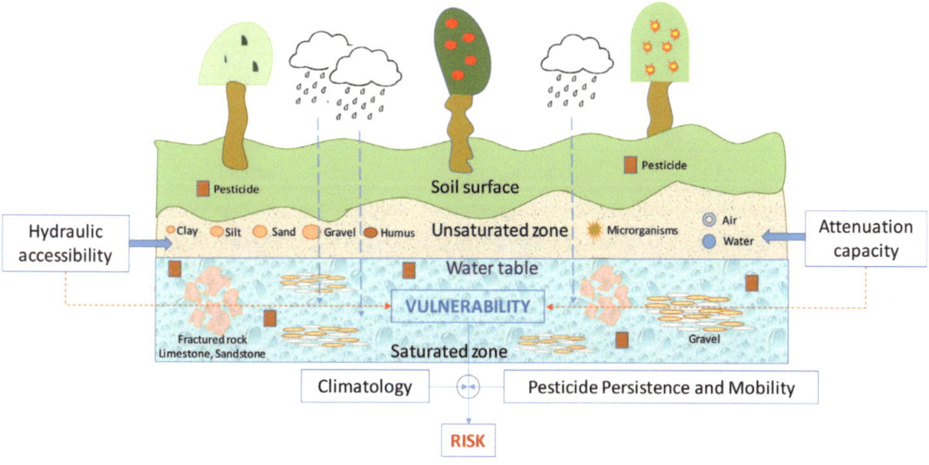

Figure 6.
Schematic drawing of the properties and factors affecting groundwater vulnerability.

water quality data to spatial variables, and (iii) overlay and index, involving obtaining and combining maps of the parameters that affect the transport of contaminants from the surface to groundwater.

Different groundwater models such as MODFLOW (1984), DRASTIC (1987), GOD (1987), AVI (1993), SINTACS (1994), SEEPAGE (1996), EPIK (1999), HAZARD-PATHWAY-TARGET (2002), INDICATOR KRIGING (2002), GLA & PI (2005), ISIS (2007) GSFLOW (2008), GWM-2005 (2009), or VULPES (2015) among others have been used to evaluate groundwater vulnerability, although these models require significant input data to run, and for most users, it is not easy to use them [99–103]. The most commonly used model is DRASTIC in the framework of GIS environment (GIS-based DRASTIC model), an overlay and index method developed by US EPA [104]. GIS is a system of hardware and software used for storage, retrieval, mapping, and analysis of geographic data showing one of the leading tools in the field of hydrogeological science that helps in assessing, monitoring, and conserving groundwater resources, while DRASTIC provides a basis for evaluating the vulnerability to pollution of groundwater resources based on hydrogeological parameters. The DRASTIC model uses seven environmental parameters (**D**epth to water, net **R**echarge, **A**quifer media, **S**oil media, **T**opography, **I**mpact of the vadose zone, and hydraulic **C**onductivity) to characterize the hydrogeological setting and evaluate aquifer vulnerability, which helps prioritize areas with respect to groundwater contamination vulnerability. Each parameter has assigned a rate and a weight (**Table 3**).

$$\textbf{DRASTIC index} = (Dr \times Dw) + (Rr \times Rw) + (Ar \times Aw) + (Sr \times Sw) + (Tr \times Tw) + (Ir \times Iw) + (Cr \times Cw) \quad (7)$$

Ratings				Weights	
Topography (% slope)		Impact of the vadose zone media			
Range	Rating	Media type	Rating	Parameter	Weight
0–2	10	Confining layer	1	Depth to water	5
2–6	9	Silt/clay	3	Net recharge	4
6–12	5	Shale	3	Aquifer media	3
12–18 3	3	Limestone	6	Soil media	2
>18	1	Sandstone	6	Topography	1
		Bedded limestone/ sandstone/shale	6	Impact of the vadose zone media	5
		Sand and gravel with clay/silt	6	Hydraulic conductivity of the aquifer	3
		Metamorphic/igneous	4	Net recharge	4
		Sand and gravel	8		
		Basalt	9		
		Karst limestone	10		

Table 3
Ratings and weights of each parameter in DRASTIC index.

where r is the rating for the parameter and w is an assigned weight for each parameter.

Thus, according to them, the governing equation becomes:

$$\textbf{DRASTIC index} = 5Dr + 4Rr + 3Ar + 2Sr + Tr + 5Ir + 3Cr \tag{8}$$

Depending on this model, five categories for groundwater vulnerability are established: very low, low, moderate, high, and very high. Two DRASTIC models (Pesticide DRASTIC GIS-based models) have been developed to predict generic groundwater vulnerability and pesticide groundwater vulnerability. They differ in weights, which are used as key factors to determine the DRASTIC vulnerability index. In the last decade, several authors have used this model to study the effect of different pesticides to groundwater vulnerability [105–107]. In other cases, a Bayesian methodology has been used to calculate the vulnerability of groundwater to pesticide contamination directly from monitoring data [108]. In this regard, passive samplers like polar organic chemical integrative samplers (POCIS) have shown to be suitable for the monitoring of pesticides with a wide range of physical-chemical properties in groundwater [109]. Many monitoring studies carried out worldwide in different countries of all continents have demonstrated the occurrence of pesticide residues in groundwater since the beginning of the actual century [18, 19, 110–112]. Among others, herbicides such as triazines (atrazine, simazine, terbuthylazine, propazine, cyanazine, terbutryn, prometryn), phenylureas (diuron, linuron, isoproturon, chlortoluron) and anilides (alachlor, acetochlor, metolachlor) and insecticides such as organophosphorus (malathion, chlorfenvinphos, dimethoate, parathion-methyl, azinphos-ethyl, chlorpyrifos, fenitrothion) and organochlorine (lindane and DDTs) and some of its transformation products (metabolites) are the most common pesticides found in groundwater.

5. Conclusions

Pesticides have important benefits in crop protection because they combat a variety of pests and diseases that could destroy crops increasing the quality of the harvested products. However, due to the heavy use of phytosanitary products (the worldwide consumption of pesticides reached 4.1 millions of tons of active ingredients in 2016), the occurrence of pesticide residues in the groundwater resources (water located beneath the soil's surface) constitutes a global problem worldwide, especially in the least developed countries where the use of plant protection products is very high. Herbicides, mainly triazine and urea compounds, have been the most detected pesticides since the beginning of this century. The pollution of soil and water bodies by pesticides used in agriculture can pose an important threat to aquatic ecosystems and drinking water resources because groundwater is the largest body of fresh water in many areas of the world. Diffuse pesticide input paths into groundwater are caused by leaching through the soil and unsaturated zone and infiltration through riverbanks and riverbeds. Therefore, the groundwater resources are vulnerable to pollution, which indicates the sensitivity of groundwater to an alteration in its quality caused by human activities. Adsorption, degradation, and movement processes are key processes to know the persistence of a pesticide and its ability to contaminate groundwater bodies. The main factors affecting the fate of pesticides are their physicochemical properties (water solubility, vapor pressure, adsorption coefficient, etc.), soil characteristics (texture, organic matter content, etc.), site (hydrogeological conditions), and management practices (method of application and dosage).

Abbreviations

CE	capillary electrophoresis
CEC	cation exchange capacity
DFOP	double first order in parallel equation
DOM	dissolved organic matter
DRASTIC	**D**epth to Water, Net **R**echarge, **A**quifer **M**edia, Soil Media, **T**opography, **I**mpact of Vadose Zone, and Hydraulic **C**onductivity
DTK	depth, half-life ($t_{1/2}$), and organic carbon normalized sorption coefficient (K_{OC})
EDs	endocrine disruptors
EFSA	European Food Safety Agency
EM	electrochemical method
EPA	Environmental Protection Agency
EPIK	development of the **E**pikarst, Effectiveness of the **P**rotective Cover, Conditions of **I**nfiltration, Development of the **K**arst Network
EU	European Union
FAO	Food Agriculture Organization
FOCUS	**FO**rum for the **C**oordination of Pesticide Fate Models and their **US**e
FOMC	First-Order Multi-Compartment
GC	gas chromatography
GIS	Geographic Information System
GLA&PI	**G**eologische **LA**ndesamter, **P**rotection Cover, **I**nfiltration Conditions
GOD	**G**roundwater Occurrence, **O**verlying Lithology and **D**epth to the Aquifer
GSFLOW	Coupled **G**round-Water and **S**urface-Water **FLOW** Model
GWM-2005	**G**round**W**ater **M**anagement process for MODFLOW-2005
ICMs	immunochemical methods
IMS	ion mobility spectrometry
IPM	Integrated Pest Management
LC	liquid chromatography
MODFLOW	**MOD**ular three-dimensional finite-difference groundwater **FLOW** model
MS	mass spectrometry
OA	organic amendment
OC	organic carbon
OECE	Organization for Economic Co-operation and Development
OPP	Office of Pesticide Programs
OM	organic matter
PMRA	Pesticide Management Regulatory Authority
POCIS	Polar Organic Chemical Integrative Samplers
PPDB	Pesticide Properties Database
PPP	plant protection product
PRZM-GW	pesticide root zone model-groundwater
SCI-GROW	screening concentration in groundwater
SEEPAGE	**S**ystem for **E**arly **E**valuation of **P**ollution **P**otential of **A**gricultural **G**roundwater **E**nvironments
SFO	simple first order

SINTACS	depth to water (**S**), net infiltration (**I**), unsaturated zone (**N**), soil media (**T**), aquifer media (**A**), hydraulic conductivity (**C**), slope (**S**)
SNV	specific numerical value
SOM	soil organic matter
US EPA	United States Environmental Protection Agency
USA	United States of America
USDA	United States Department of Agriculture
VULPES	**VUL**nerability to **PES**ticides

Author details

Gabriel Pérez-Lucas[1], Nuria Vela[2], Abderrazak El Aatik[2] and Simón Navarro[1]*

1 Faculty of Chemistry, Department of Agricultural Chemistry, Geology and Pedology, University of Murcia, Murcia, Spain

2 Faculty of Health Science, Applied Technology Group to Environmental Health, Catholic University of Murcia, Murcia, Spain

*Address all correspondence to: snavarro@um.es

References

[1] EUROSTAT. The European Union in the World. Luxemburg: The European Commission; 2018. Available from: http://ec.europa.eu/eurostat. [Accessed: 20 Sep. 2018]

[2] FAOSTAT. Food and Agriculture Organization (FAO) of the United Nations [Internet]. 2018. Available from: ftp://ftp.fao.org/FAOSTAT [Accessed: 19 Sep. 2018]

[3] Food and Agriculture Organization (FAO) of the United Nations. International Code of Conduct on the Distribution and Use of Pesticides. 2003. Available from: ftp://ftp.fao.org/docrep/fao/005/y4544e/y4544e00.pdf [Accessed: 12 Sep. 2018]

[4] Regulation (EC) No 1107/2009 of the European Parliament and of the Council of 21 October 2009 concerning the placing of plant protection products on the mark. Official Journal of the European Union. 2009;**L 309**:1-50

[5] Shen L, Wania F, Lei YD, Teixeira C, Muir DCG, Bidleman TF. Atmospheric distribution and long-range transport behaviour of organochlorine pesticides in North America. Environmental Science & Technology. 2005;**39**:409-420

[6] Van der Werf HMG. Assessing the impact of pesticides on the environment. Agriculture, Ecosystems & Environment. 1996;**60**:81-96

[7] International Programme on Chemical Safety (IPCS). Global Assessment of the State-of-the-Science of Endocrine Disruptors. Geneva, Switzerland: World Health Organization; 2002

[8] Mnif W, Hassine AIH, Bouaziz A, Bartegi A, Thomas O, Roig B. Effect of endocrine disruptor pesticides: A review. International Journal of Environmental Research and Public Health. 2011;**8**:2265-2303

[9] Roig B, Mnif W, Hassine AI, Zidi I, Bayle S, Bartegi A, et al. Endocrine disrupting chemicals and human risk assessment: A critical review. Critical Reviews in Environmental Science and Technology. 2013;**43**:2297-2351

[10] Giulivo M, López de Alda M, Capri E, Barceló D. Human exposure to endocrine disrupting compounds: Their role in reproductive systems, metabolic syndrome and breast cancer. A review. Environmental Research. 2016;**151**: 251-264

[11] Diamanti-Kandarakis E, Bourguignon JP, Giudice LC, Hauser R, Prins GS, Soto AM, et al. Endocrine disrupting chemicals: An Endocrine Society scientific statement. Endocrine Reviews. 2009;**30**:293-342

[12] European Environment Agency (EEA). The Impacts of Endocrine Disrupters on Wildlife, People and Their Environments—The Weybridge +15 (1996–2011) Report. (Technical Report No. 2/2012). Copenhagen, Denmark: European Environment Agency; 2012. 112p

[13] Directive 2006/118/EC of the European Parliament and of the Council of 12 December 2006 on the protection of groundwater against pollution and deterioration. Official Journal of the European Union. 2006;**L 372**:19-31

[14] Reichenberger S, Bach M, Skitschak A, Frede HG. Mitigation strategies to reduce pesticide inputs into the ground- and surface water and their effectiveness. A review. Science of the Total Environment. 2007;**384**:1-35

[15] Arias-Estévez M, López-Periago E, Martínez-Carballo E, Simal-Gándara J, Mejuto JC, García-Río L. The mobility and degradation of pesticides in soils and the pollution of groundwater

resources. Agriculture, Ecosystems & Environment. 2008;**123**:247-260

[16] Gurdak JJ. Groundwater vulnerability. In: Slamian S, editor. Handbook of Engineering Hydrology, Environmental Hydrology and Water Management. Boca Raton, FL: CRC-Press; 2014. pp. 145-162

[17] National Research Council. Ground Water Vulnerability Assessment, Predicting Relative Contamination Potential under Conditions of Uncertainty. Washington, DC: National Academy Press; 1993. 210p

[18] Mouvet C. Pesticides in European groundwaters: Biogeochemical processes, contamination status and results from a case study. In: Quevauviller P, editor. Groundwater Science and Policy: An International Overview. Paris, France; 2007. pp. 545-583

[19] Toccalino PL, Gilliom RJ, Lindsey BD, Rupert MG. Pesticides in groundwater of the United States: Decadal-scale changes, 1993–2011. Groundwater. 2014;**52**:112-125

[20] Directive 2009/128/EC of the European Parliament and of the Council of 21 October 2009 establishing a framework for Community action to achieve the sustainable use of pesticides. Official Journal of the European Union. 2009;**L 309**:71-86

[21] Food and Agricultural Organisation of the United Nations. Integrated Pest Management. 2012. Available from: http://www.fao.org/agriculture/crops/core-themes/theme/pests/ipm [Accessed: 25 Sep. 2018]

[22] United States Environmental Protection Agency Pesticides and Food. What Integrated Pest Management Means. 2011. Available from: http://www.epa.gov/pesticides/food/ipm.htm [Accessed: 25 Sep. 2018]

[23] Ehi-Eromosele CO, Nwinyi OC, Ajani OO. Integrated pest management. In: Weed and Pest Control—Conventional and New Challenges. Rijeka: InTech; 2015. pp. 105-115

[24] Soil Science Division Staff. Soil survey manual. In: Ditzler C, Scheffe K, Monger HC, editors. USDA Handbook 18. Washington, D.C, USA: Government Printing Office; 2017

[25] Porta J, López-Acevedo M, Roquero C. Edafología para la Agricultura y el Medio Ambiente. 3ª Edición. Ediciones Mundi-Prensa, Madrid, España; 2003

[26] Navarro G, Navarro S. Química Agrícola. 3ª Edición. Ediciones Mundi-Prensa. Madrid, España; 2013

[27] Navarro S, Vela N, Navarro G. An overview on the environmental behaviour of pesticide residues in soils. Spanish Journal of Agricultural Research. 2007;**5**:357-375

[28] Tadeo JL, Pérez RA, Albero B, Garcia-Valcarcel AI, Sanchez-Brunete C. Review of sample preparation techniques for the analysis of pesticide residues in soil. Journal of Association of Official Agricultural Chemists (AOAC) International. 2012;**95**:1258-1271

[29] Semen S, Mercan S, Acikkol M. A general overview of pesticides in soil: Requirement of sensitive and current residue analysis methods. In: Kars H, Van Den Eijkel L, editors. Soil in Criminal and Environmental Forensics. London/Berlin, Holland: Springer; 2016. pp. 163-180

[30] Durães N, Novo LAB, Candeias C, da Silva EF. Distribution, transport and fate of pollutants. In: Duarte AC, Cahada A, Rocha-Santos T, editors. Soil Pollution: From Monitoring to Remediation. London, United Kingdom: Academic Press, an imprint of Elsevier; 2018. pp. 29-56

[31] Børgensen CD, Fomsgaard IS, Plauborg F, Schelde K, Spliid NH. Fate of Pesticides in Agricultural Soils. Tjele, Denmark: Danish Centre for Food and Agriculture; 2015

[32] Gavrilescu M. Fate of pesticides in the environment and its bioremediation. Engineering in Life Sciences. 2005;**5**: 497-526

[33] Organization for Economic Cooperation and Development (OECD). Guidelines for Testing of Chemicals, No 106, Adsorption-Desorption using a Batch Equilibrium Method. Paris: Organization for Economic Cooperation and Development (OECD); 2000

[34] Organization for Economic Cooperation and Development (OECD). Guidelines for Testing of Chemicals, No 307, Aerobic and Anaerobic Transformation in Soil. Paris: Organization for Economic Cooperation and Development (OECD); 2002

[35] Organization for Economic Cooperation and Development (OECD). Guidelines for Testing of Chemicals, No 312, Leaching in Soil Columns. Paris: Organization for Economic Cooperation and Development (OECD); 2007

[36] Roof AMM. Basic principles of environmental photochemistry. Part B. In: Hutzinger O, editor. The Handbook of Environmental Chemistry. Vol. 2. Berlin: Springer-Verlag; 1982. pp. 1-17

[37] Das S, editor. Microbial Biodegradation and Bioremediation. Amsterdam, Netherlands: Elsevier; 2014. 642p

[38] Gevao B, Semple KT, Jones KC. Bound pesticide residues in soils. A review. Environmental Pollution. 2000; **108**:3-14

[39] Roberts TR. Non-extractable pesticide residues in soils and plants. Pure and Applied Chemistry. 1984;**56**:945-956

[40] Führ F, Ophoff H, Burauel P, Wanner U, Haider K. Modification of the definition of bound residues. In: Pesticide Bound Residues - Workshop-Senate Commission for the Assessment of Chemicals Used in Agriculture. Weinheim, Germany: Wiley-VCH Verlag; 1998. pp. 175-176

[41] Boesten JJTI, Aden K, Beigel CY, Beulke S, Dust M, Dyson JS, et al. Guidance Document on Estimating Persistence and Degradation Kinetics from Environmental Fate Studies on Pesticides in EU Registration. Report of the FOCUS (Forum for the Co-ordination of pesticide fate models and their Use) Work Group on Degradation Kinetics, EC Document Reference Sanco/10058/2005 version 2.0; 2005

[42] Cohen SZ, Wauchope RD, Klein AW, Eadsforth CV, Graney R. Offsite transport of pesticides in water: Mathematical models of pesticide leaching and runoff. Pure and Applied Chemistry. 1995;**12**:2109-2148

[43] Steffens K, Larsbo M, Moeys J, Jarvis N, Lewan E. Predicting pesticide leaching under climate change: Importance of model structure and parameter uncertainty. Agriculture, Ecosystems and Environment. 2013;**172**: 24-34

[44] Langeron J, Blondel A, Sayen S, Hénon E, Couderchet M, Guillon E. Molecular properties affecting the adsorption coefficient of pesticides from various chemical families. Environmental Science and Pollution Research. 2014;**21**:9727-9741

[45] Agriculture & Environment Research Unit (AERU) at the University of Hertfordshire. 2018. Pesticide Properties Database. Available from: https://sitem.herts.ac.uk/aeru/footprint/es/ [Accessed: 27 Sep. 2018]

[46] Karickhoff SW, Brown DS, Scott TA. Sorption of hydrophobic pollutants

on natural sediments. Water Research. 1979;**13**:421-428

[47] Hermosín MC, Cornejo J. Role of soil clay fraction in pesticide adsorption. Defining a Kclay. In: Copin A, Housins G, Pussemier L, Salembier JF, editors. Environmental Behaviour of Pesticides and Regulatory Aspects. London, UK: Royal Society of Chemistry; 1994. pp. 206-207

[48] Rodríguez-Cruz MS, Sánchez-Martín MJ, Andrades MS, Sánchez-Camazano M. Modification of clay barriers with a cationic surfactant to improve the retention of pesticides in soils. Journal of Hazardous Materials. 2009;**B139**:363-372

[49] Rodríguez-Cruz MS, Sánchez-Martín MJ, Andrades MS, Sánchez-Camazano M. Retention of pesticides in soil columns modified in situ and ex situ with a cationic surfactant. Science of the Total Environment. 2007;**378**:104-108

[50] Papiernik S, Koskinen W, Cox-Rice P, Clay S, Werdin-Pfisterer N, Norberg K. Sorption-desorption of imidacloprid and its metabolites in soil and vadose zone materials. Journal of Agricultural and Food Chemistry. 2006;**54**: 8163-8170

[51] Sadegh-Zadeh F, Wahid SA, Jalili B. Sorption, degradation and leaching of pesticides in soils amended with organic matter: A review. Advances in Environmental Technology. 2017;**2**: 119-132

[52] Wauchope RD, Yeh S, Linders J, Kloskowski R, Tanaka K, Rubin B, et al. Pesticide soil sorption parameters: Theory, measurement, uses, limitations and reliability. Pest Management Science. 2002;**58**:419-445

[53] Bastida F, Kabdeler E, Moreno JL, Ros M, García C, Hernández T. Application of fresh and composted organic wastes modifies structure, size and activity of soil microbial community under semiarid climate. Applied Soil Ecology. 2008;**40**:318-329

[54] Cox L, Celis R, Hermosin MC, Becker A, Cornejo J. Porosity and herbicide leaching in soils amended with olive-mill wastewater. Agriculture, Ecosystems & Environment. 1997;**65**: 151-161

[55] Celis R, Barriuso E, Houot S. Effect of liquid sewage sludge addition on atrazine sorption and desorption by soil. Chemosphere. 1998;**37**:1091-1107

[56] Majumdar K, Singh N. Effect of soil amendments on sorption and mobility of metribuzin in soils. Chemosphere. 2007;**66**:630-637

[57] Cabrera A, Cox L, Koskinen WC, Sadowsky MJ. Availability of triazine herbicides in aged soils amended with olive oil mill waste. Journal of Agricultural and Food Chemistry. 2008; **56**:4112-4119

[58] Delgado-Moreno PA. Compost and vermicompost of olive cake to bioremediate triazines-contaminated soil. Science of the Total Environment. 2009;**407**:1489-1495

[59] Jiang L, Ma L, Sui Y, Han SQ, Yang H. Mobilization and plant accumulation of prometryne in soil by two different sources of organic matter. Journal of Environmental Monitoring. 2011;**13**: 1935-1943

[60] Grenni P, Rodríguez-Cruz MS, Herrero-Hernandez E, Marin-Benito JM, Sanchez-Martin MJ, Caracciolo AB. Effects of wood amendments on the degradation of terbuthylazine and on soil microbial community activity in a clay loam soil. Water, Air, & Soil Pollution. 2012;**223**:5401-5412

[61] Gámiz B, Celis R, Cox L, Hermosín MC, Cornejo J. Effect of olive-mill waste addition to soil on sorption, persistence,

and mobility of herbicides used in Mediterranean olive groves. Science of the Total Environment. 2012;**429**: 292-299

[62] Fenoll J, Vela N, Navarro G, Pérez-Lucas G, Navarro S. Assessment of agro-industrial and composted organic wastes for reducing the potential leaching of triazine herbicide residues through the soil. Science of the Total Environment. 2014;**493**:124-132

[63] Fenoll J, Garrido I, Hellín P, Flores P, Vela N, Navarro S. Use of different organic wastes as strategy to mitigate the leaching potential of phenylurea herbicides through the soil. Environmental Science and Pollution Research. 2015;**22**:4336-4349

[64] Marín-Benito JM, Sánchez-Martín MJ, Rodríguez-Cruz MS. Impact of spent mushroom substrates on the fate of pesticides in soil, and their use for preventing and/or controlling soil and water contamination: A review. Toxics. 2016;**4**:1-24

[65] Joseph SD, Camps-Arbestain M, Lin Y, Munroe P, Chia CH, Hook J, et al. An investigation into the reactions of biochar in soil. Australian Journal of Soil Research. 2010;**48**:501-515

[66] Lone AH, Najar GR, Ganie MA, Sofi JA, Ali T. Biochar for sustainable soil health: A review of prospects and concerns. Pedosphere. 2015;**25**:639-653

[67] Khorram MS, Zhang Q, Lin D, Zheng Y, Fang H, Yu Y. Biochar: A review of its impact on pesticide behavior in soil environments and its potential applications. Journal of Environmental Sciences. 2016;**44**: 269-279

[68] Shareef TME, Zhao B. Review paper: The fundamentals of biochar as a soil amendment tool and management in agriculture scope: An overview for farmers and gardeners. Journal of Agricultural Chemistry and Environment. 2017;**6**:38-61

[69] Worral F, Fernández-Pérez M, Johnson AC, Flores-Cesperedes F, González-Pradas E. Limitations on the role of incorporated organic matter in reducing pesticide leaching. Journal of Contaminant Hydrology. 2001;**49**: 241-262

[70] Chantigny MH. Dissolved and water-extractable organic matter in soils: A review on the influence of land use and management practices. Geoderma. 2003;**113**:357-380

[71] Huang X, Lee S. Effects of dissolved organic matter from animal waste effluent on chlorpyrifos sorption by soils. Journal of Environmental Quality. 2001;**30**:1258-1265

[72] Velarde L, Cabrera P, Hermosín MC, Cornejo J. Dissolved organic carbon interactions with sorption and leaching of diuron in organic-amended soils. European Journal of Soil Science. 2007; **58**:714-721

[73] Spark K, Swift R. Effects of soil composition and dissolved organic matter on pesticide sorption. Science of the Total Environment. 2002;**298**: 147-161

[74] Briceño G, Palma G, Durán N. Influence of organic amendment on the biodegradation and movement of pesticides. Critical Reviews in Environmental Science and Technology. 2007;**37**:233-271

[75] Katagi T. Soil column leaching of pesticides. Reviews of Environmental Contamination and Toxicology. 2013; **221**:1-105

[76] Environmental Protection Agency (EPA). Fate, transport and transformation test guidelines. Leaching studies. OPPTS 835.1240, EPA 712-C-08-010.

Washington DC, USA: Environmental Protection Agency; 2008

[77] Banzahaf S, Hebig KH. Use of column experiments to investigate the fate of micropollutants. A review. Hydrology and Earth System Sciences. 2016;**20**:3719-3737

[78] Kordel W, Klein M. Prediction of leaching and groundwater contamination by pesticides. Pure and Applied Chemistry. 2006;**78**:1081-1090

[79] Hamaker J. The interpretation of soil leaching experiments. In: Haque R, Freed V, editors. Environmental Dynamics of Pesticides. New York: Plenum Press; 1975. pp. 115-133

[80] McCall PJ, Laskowski DE, Swann RL, Disburger HJ. Measurement of sorption coefficients of organic chemicals and their use, in environment fate analysis. In: Test Protocols for Environmental Fate and Movement of Toxicants. Proceeding Symposium Association of Analytical Chemists, 94th Annual Meeting, AOAC, Washington DC. 1981. pp. 89-109

[81] Briggs GG. Theoretical and experimental relationships between soil adsorption, octanol-water partition coefficient, water solubilities, bioconcentration factors, and parachor. Journal of Agricultural and Food Chemistry. 1981;**29**:1050-1059

[82] Laskowski DA, Goring CAI, McCall PJ, Swann RL. Terrestrial environment. In: Conway RA, editor. Environmental Risk Analysis for Chemicals. NY: Van Nostrand Reinhold Co; 1982. pp. 198-240

[83] Cohen SZ, Creeger SM, Carsel RF, Enfield CG. Potential for pesticide contamination of groundwater resulting from agricultural uses. In: Krueger RF, Seiber JN, editors. Treatment and Disposal of Pesticide Wastes. ACS Symposium. Vol. 259. Washington DC: American Chemical Society; 1984. pp. 297-325

[84] Hornsby AG. Site-specific pesticide recommendations: The final step in environmental. Weed Technology. 1992;**6**:736-742

[85] Rao PSC, Hornsby AG, Jessup RE. Indices for ranking the potential for pesticide contamination of groundwater. Proceedings of the Soil and Crop Science Society of Florida. 1985;**44**:1-8

[86] Gustafson D. Groundwater ubiquity score: A simple method for assessing pesticide leachability. Environmental Toxicology and Chemistry. 1989;**8**: 339-357

[87] Meeks YJ, Dean JD. Evaluating groundwater vulnerability to pesticides. Journal of Water Resources Planning and Management. 1990;**116**:693-707

[88] Johnson B. Setting Revised Specific Numerical Values. California Environmental Hazards Assessment Program Report EH-91-6. Sacramento, CA. USA: Environmental Protection Agency; 1991

[89] Warren RL, Weber JB. Evaluating pesticide movement in North Carolina soils. Soil Science Society of America Proceedings. 1994;**37**:23-35

[90] RA ML, Weber JB, Warren RL. Soil Facts, Protecting Groundwater in North Carolina: A Pesticide and Soil Ranking System. College of Agricultural and Life Sciences. Raleigh, NC, USA: North Carolina State University, North Carolina Cooperative Extensive Service; 1994

[91] Li ZC, Yost RS, Green RE. Incorporating uncertainty in a chemical leaching assessment. Journal of Contaminant Hydrology. 1998;**29**:285-299

[92] Spadotto CA. Screening method for assessing pesticide leaching potential.

Pesticidas: Revista de Ecotoxicologia E Meio Ambiente. 2002;**12**:69-78

[93] Gramatica P, Di Guardo A. Screening of pesticides for environmental partitioning tendency. Chemosphere. 2002;**47**:947-956

[94] Papa E, Castiglioni S, Gramatica P, Nikolayenko V, Kayumov O, Calamari D. Screening the leaching tendency of pesticides applied in the Amu Darya Basin (Uzbekistan). Water Research. 2004;**38**:3485-3494

[95] Webb RMT, Wieczorek ME, Nolan BT, Hancock TC, Sandstrom MW, Barbash JE, et al. Variations in pesticide leaching related to land use, pesticide properties, and unsaturated zone thickness. Journal of Environmental Quality. 2008;**37**:1145-1157

[96] The Groundwater Foundation, Lincoln, NE, USA. What is Groundwater? 2018. Available from: http://www.groundwater.org/get-informed/basics/groundwater.html [Accessed: 4 Oct, 2018]

[97] Barrett M. Initial Tier Screening of Pesticides for Groundwater Concentration Using the SCI-GROW Model. Washington, D.C.: U.S. Environmental Protection Agency; 1997

[98] Brady D. Approval of PRZM-GW for Use in Drinking Water Exposure Assessments. U.S: Environmental Protection Agency; 2012

[99] McDonald MG, Harbaugh AW. A modular three-dimensional finite-difference groundwater flow model. US Geological Survey. Open-File; 1984. Report 83-875

[100] Markstrom SL, Niswonger RG, Regan RS, Prudic DE, Barlow PM. GSFLOW—Coupled Ground-Water and Surface-Water Flow Model Based on the Integration of the Precipitation-Runoff Modeling System (PRMS) and the Modular Ground-Water Flow Model (MODFLOW-2005). US Geological Survey. Techniques and Methods 6-D1; 2008. 240p

[101] Ahlfeld DP, Baker KM, Barlow PM. GWM-2005—A Groundwater-Management Process for MODFLOW-2005 with Local Grid Refinement (LGR) Capability. US Geological Survey. Techniques and Methods 6-A33; 2009. 65p

[102] Kumar P, Bansod BKS, Debnath SK, Thakur PK, Ghanshyam C. Index-based groundwater vulnerability mapping models using hydrogeological settings: A critical evaluation. Environmental Impact Assessment Review. 2015;**51**:38-49

[103] Di Guardo A, Finizio A. A client-server software for the identification of groundwater vulnerability to pesticides at regional level. Science of the Total Environment. 2015;**530**:247-256

[104] Aller L, Bennett T, Lehr J, Petty R. DRASTIC: A Standardized System for Evaluating Ground Water Pollution Potential using Hydrogeologic Settings. Washington, D.C., EPA/600/2-85/018: U.S. Environmental Protection Agency; 2004

[105] Ahmed AA. Using generic and pesticide DRASTIC GIS-based models for vulnerability assessment of the quaternary aquifer at Sohag, Egypt. Hydrogeology Journal. 2009;**17**: 1203-1217

[106] Anane M, Abidi B, Lachaal F, Limam A, Jellali S. GIS-based DRASTIC, pesticide DRASTIC and the susceptibility index (SI): Comparative study for evaluation of pollution potential in the Nabeul-Hammamet shallow aquifer, Tunisia. Hydrogeology Journal. 2013;**21**:715-731

[107] Saha D, Alam F. Groundwater vulnerability assessment using

DRASTIC and Pesticide DRASTIC models in intense agriculture area of the Gangetic plains, India. Environmental Monitoring and Assessment. 2014;**186**: 8741-8763

[108] Worrall F, Besien T. The vulnerability of groundwater to pesticide contamination estimated directly from observations of presence or absence in wells. Journal of Hydrology. 2004;**303**:92-107

[109] Ahrens L, Daneshvar A, Lau AE, Kreuger J. Characterization of five passive sampling devices for monitoring of pesticides in water. Journal of Chromatography A. 2015;**1405**:1-11

[110] Ouedraogo I, Defourny P, Vanclooster M. Mapping the groundwater vulnerability for pollution at the pan African scale. Science of the Total Environment. 2016;**544**:939-953

[111] Huang PM, Iskandar IK. Soils and Groundwater Pollution and Remediation: Asia, Africa, and Oceania. Boca Raton, FL, USA: Lewis Publishers; 1999. 386p

[112] Sundaram B, Coram J. Groundwater Quality in Australia and New Zealand: A Literature Review. The Guidelines for Groundwater Protection in Australia. Final Report. Geoscience Australia; 2009. 37p

Chapter 4

Pesticides, Anthropogenic Activities, and the Health of Our Environment Safety

Mona Saud AL-Ahmadi

Abstract

Mankind depends on agricultural products for food consumption. Increasing population (more than 7 billion) requires significant growth in crop yield to meet essential demand. This aim was achieved through the use of pesticides to protect crops from diseases. Pesticides are toxic by design for organisms that can threaten food products. Their mode of action is by targeting systems or enzymes in the pests that may be similar to human system and therefore pose risks to human health and the environment as well. The WHO recommended classifying pesticides according to their toxicity and chemicals according to their chronic health and environmental hazards.

Keywords: pesticides, classification of pesticides, pesticide hazards, future of pesticides

1. Introduction

Agriculture is the primary source for human food; it provides different kinds of crop production. Most common crops include wheat, rice, corn, beans, different vegetables, and season fruits.

In 2015, 7.4 billion people call earth their home. Population is projected to reach 9.7 billion by 2050 and 11.2 billion by 2100 [1]. Optimizing crop yields becomes even a more critical factor affecting the availability and affordability of food to meet increasing population demand.

Plant diseases are major factors that affect crop production. Al-Sadi [2] reported that plant diseases can affect plants by interfering with several processes such as the absorbance and translocation of water and nutrients, photosynthesis, and flower and fruit development. Infection of plants by pathogens can have serious consequences on plant health which consequently affect human health. Viruses, bacteria, and fungi that infect plants do not usually cause infection in humans [3]. The ultimate consequences of different plant diseases are reduction in crop production, reducing food availability which may lead starvation in some areas. The famous incident of plant pathology is potato disease caused by *Phytophthora infestans* fungi, which destroyed potatoes that were the main crop in Ireland during 1845–1850, where about 1 million people died and another million immigrated to other countries [4].

This disaster and other similar ones resulted from plant pathogens. Minimizing this risk requires efficient methods and practices to control pests (insects, bacteria, fungi, viruses, etc.). The term "pesticide" indicates any substance or mixture of substance used to kill, "repel," or otherwise control a "pest," including insects, snails, rodents, fungi, bacteria, and weeds [5].

The early methods used were simple and depended on traditional ways in specific places; however, these traditional practices were insufficient to control pests efficiently. Improvement of pest control gradually started to show satisfactory results for farmers and food manufacturing by the nineteenth century through the introduction of two natural pesticides (pyrethrum and rodent). In 1939 Muller discovered that DDT was a very effective insecticide and quickly became the most widely used pesticide in the world. Not until the 1960s when the harmful side effects of the application of DDT was discovered [6]. Despite the harmful effects of DDT, demand for pesticides continued to increase throughout the world. This is due to the many benefits attributed to pesticides; the most obvious benefits are economic, protection of commodity yield and quality, and the reduction of other costly inputs such as labor and fuel [7]. Pesticides play an essential role in farm profitability, providing reliable supplies of agricultural product, improving the quality of the product [8]. Notwithstanding pesticide benefits, there is plenty of evidence of both direct and indirect dangers involved in the use of these chemical substances both for humans and the environment [8].

For example, the contamination of pesticides may happen in several ways during manufacturing, storing, shipping, application in fields, warehouses, and wrong use by peoples. Several accidents have occurred in different parts of the world: India (1986), Italy (1976), Germany (1953), and Ethiopia (2017) [9]. Maksymiv [10] grouped the side effects of the excessive use of pesticides into diseases of ecospecies such as erosion, loss of soil fertility, pollution of water system, and biological community impact including loss of crop, animal genetic resources, elimination of natural enemies, genetic resistance to pesticides, contamination, and changes to natural control mechanisms.

Due the undesirable side effects of synthetic pesticides, search for safer analogue pesticides of natural origin is one of the most important goals. Potential alternatives to pesticides are available and include specific methods of plant cultivation, use of biological pest control, plant genetic engineering, and methods of interfering with insect breeding [11]. The most common alternative to synthetic pesticide pest control is biopesticides.

Biopesticides are a certain type of pesticides derived from natural material such as animal, plants, bacteria, and certain minerals. As of April 2016, there were 299 registered biopesticide active ingredients and 1401 active biopesticide product registration (US Environmental Protection Agency EPA).

Biopesticide offers a more sustainable solution to pest control than synthetic alternative. Botanical pesticides do not present the residue problems [12]. Microbial pesticides contain a microorganism as the active ingredient; they can control many different kinds of pest, although each separate active ingredient is relatively specific for its target pest(s). Biochemical pesticides are naturally occurring substance that interferes with growth or mating such as plant growth regulators or substances that repel or attract pests, such as pheromone [13]. Even with the satisfactory results of biopesticides on pest management, the efficacy at different geographical conditions and slow pest control makes them less desirable by farmers [14]. The science of biopesticides is still considered to be young and evolving. Some of the biopesticides are under development; this may prove to be excellent alternatives to chemical pesticides. Further research is needed in several areas such as production, formulation, delivery, and commercialization of the products [15].

In recent years, a new technology that provided a sustainable solution is nanotechnology through the development of nanopesticides for conventional agricultural use [16, 17]. Nanopesticides are small engineered structures that provide pesticide properties or formulation of active ingredient of pesticides in nanoform; these nanostructures show slow degradation and controlled release of nanopesticides which make them environmentally safer and less toxic compared to chemical pesticides [14]. The nanosystems have shown great capability of controlled release pattern of active ingredient (AI) making them more efficient for long time period usability that can solve eutrophication and residual pesticide accumulation problem [18]. In spite of the excellent results of this nanotechnology in the agriculture field, further and deep studies should be conducted to ensure application safety.

2. Plant pathology and impacts of pesticide usage

A plant disease is usually defined as abnormal growth and/or dysfunction of a plant resulting from disturbance in normal life process or infections of living organisms (biotic) and nonliving environmental conditions (abiotic) [19].

Plant disease is best managed through an integrated approach, which includes a combination of:

1. Cultural management that utilize plant performance in the local climate, use of disease-resistant varieties when possible, and plant-certified seeds or seed pieces.

2. Mechanical management includes rototilling in the fall, which exposes pathogens, insect eggs, and weed seeds to cold winter temperatures. This action speeds the decomposition of crop residues, improving soil organic matter.

3. Biological control, depending on the use of compost, compost teas, and hyperparasite products that may reduce pathogens by introducing beneficial microbes. Planting flowering plants attracts beneficial insects to all stages of development.

4. Chemical control, depending on the effect of different types of pesticides to manage the problem; this will be efficient after identifying the cause of a plant problem first and applying it in the correct time using the recommended method [19].

It has been reported that field losses from pest's average 35% from the world's main food crops. Direct yield losses caused by pathogens, animals, and weeds are altogether responsible for losses ranging between 20 and 40% of global agricultural productivity [20–22].

Although weeds are the major cause of crop loss on a global scale, significant losses are suffered by agricultural crops due to insect damage and plant diseases; estimated worldwide annual production tonnage (%) age loses attributed to pests at the start of the twenty-first century are 18%, due to animal pests, 16% microbial diseases (of which 70–80% caused by fungi), and 34% weeds, totaling 68% average annual loss of crop production tonnage [22].

Oerke [23] reported that the total global potential loss due to pests are estimated at 26–30% for sugar beet, barley, soybean, wheat, and cotton and 35, 39, and 40% for maize, potatoes, and rice. Plant protection in general and protection of crops against plant diseases in particular have an obvious role to play in meeting the growing demand for food quality and quantity [24]. It involves physical, biological, and chemical methods [25].

The only way to reduce crop losses is integrated pest management. Integrated pest management (IPM) is a system approach that combines different crop protection practices with careful monitoring of pest and their natural enemies [26, 27]. The primary IPM method includes synthetic chemical pesticides that are classed by regulators as low-risk compounds and have high levels of selectivity, such as synthetic insect growth regulators.

1. Crop cultivars bred with total or partial pest resistance.

2. Cultivation practices, such as crop rotation intercropping or under sowing.

3. Physical methods, such as mechanical welders:

    - Natural products, such as semichemical or biocidal plant extracts

    - Biological control with natural enemies, including different pathogens of plants

    - Decision support tools to inform when it is economically beneficial to apply pesticides and other controls

Ghandler et al [28] reported that although pesticides act similarly despite their chemical active group. When applied to crops or directly to the soil, for example, systemic insecticides, organophosphates, and carbamates generally persist from only a few hours to several months. However, they have been fatal to large numbers of birds on turf and in agriculture and negatively impacted breeding success in birds [29]. Nanopesticides or nanoplant protection products represent an emerging technological development that, in relation to pesticide use, could offer a range of benefits including increased efficacy, durability, and a reduction in the amount of active ingredients that need to be used [30].

Biopesticides are natural products that can be considered as sufficient alternative of synthetic pesticide in pest management.

3. Pesticides in agriculture and their benefits

The farmers around the world had used different methods and ways to fight the causes that lead to reducing crop yield, most of these methods were simple and traditional, and the result were not satisfactory until the use of pesticide application started.

Pesticides include natural and synthetic substances used to control harmful pests such as insects, plant disease organisms, and weed, as well as many other living organisms that endanger the food supply, health, or comfort [8].

The word "pesticides" is a term for all insecticides, herbicides, fungicides, rodenticides, wood preservatives, garden chemicals, and household disinfectants that may be used to kill some pests [31].

A pesticide controls any pest including vectors of human or animal diseases and unwanted species of plants or animals causing harm or interfering with the production, processing, storage, or marketing of food and agricultural commodities [32].

Pesticides are a chemical group widely used by humans, both to protect the production from harmful organisms and quality of crops and for control of vectors and pests of public health [33]. In the last decade, pesticide sales have been roughly stable worldwide with an overall budget of $40 billion, with the US market accounting for 31.6% of the total [34]. In the last decade, the most significant increase in demand for pesticides has occurred in Central and South America (6.7% annual increase from 2004 to 2014). Followed by the Asian market (4% annual increase from 2004 to 2014); the latter is the second largest after North America. Even the small African market, accounting for 3.5% of global pesticide expenditure in 2004, has shown a sharp 6.4% annual increase during the same period. An annual increase has also been observed in Europe [35].

The use of pesticides in agriculture has led to significant improvement in crop yield per hectare of land [36]. The economy was boosted, crop yields were tremendously increased, and so were the decreases in fatalities insect-borne diseases [31]. Cooper and Dobson [37] demonstrate the three main effects of pesticides:

1. Controlling agriculture pests (including disease and weeds).

2. Controlling human and livestock disease vector nuisance organisms.

3. Preventing or controlling organisms that harm other human activities and structures.

The other strategy of protecting crops is to utilize biorational pesticides, such as biopesticides as alternative to synthetic chemicals. As synthetic pesticides are withdrawn owing to resistance problems or because they are no longer commercially viable, biopesticides are used as a replacement especially since they do not feature residue problems, which are a matter of significant concern for consumers. Currently, biopesticides comprise a small share of the total crop protection market globally, with a value of about $ 3 billion worldwide, accounting for just 5% of the total crop protection market [38, 39].

The most important characteristics that distinguish biopesticides are (a) short RELs (most are 4 hours), (b) zero-day preharvest intervals (PHI), (c) generally safer to plants, (d) low-risk to environmental, (e) quicker to market at lower overall cost—3 years and $5 million to develop vs. 10 years and $200 million, (f) complex modes of action [40].

Despite synthetic pesticides' significant effectiveness on pest and crop diseases, their harmful side effects on plants, soil, and the environment require safer products. Biopesticides clearly have a potential role to play in the development of future integrated pest management strategies, and it is very likely that in the future their role will be more significant in agriculture and forestry [41].

New technology that depends on nanosize of different materials started to spread around the world, because of various and efficient application results. Among nanotechnology sections, pesticides are receiving increasing interest with the development of a range of plant protection products that termed "nanopesticides" [42, 43]. Nanopesticides involve either very small practices of a pesticide active ingredient or other small engineered structures with useful pesticide properties [44]. Dubey et al. [45] reported that there is very limited knowledge about the nanoparticle's long-term adverse effects on soil, plants, and ultimately humans; an intelligent use of nanotechnology may help to achieve food security with the qualitative and sustainable environment.

4. Classification of agrochemical pesticides

4.1 Synthetic pesticides

Synthetic pesticides are classified based on various ways depending on the needs; the three most popular ways to classifying pesticides are the mode of action, the targeted pest species, and the chemical composition of the pesticides [46]. The World Health Organization (WHO) proposed a classification of synthetic pesticides based on their health risks and estimating the median lethal dose (LD50) that produces death in 50% of exposed animals [47].

Pesticide formulation includes emulsifiable concentrates (EC) which are fine suspensions of oil droplets in water and appears milky in color. Wet table powders (WP) are suspensions of fine particles suspended in water. Granules (G) are prepared by mixing the active ingredient with clay for outdoor use. Baits are obtained by mixing the active ingredient with food base especially used for the control of rodents. Dusts (D) must be applied dry and cannot be mixed with water. Fumigants are gaseous insecticides usually packaged under pressure and stored as liquids. Some are tablets or pellets that release gas when mixed with water [31].

Pesticide's mode of action can be classified as contact (non-systemic) and systemic pesticides [31].

Garcia et al. [33] describe that pesticides are classified by target organism (e.g., insecticides, herbicides, fungicides, miticides, nematicides, molluscicides, and rodenticides) and chemical structure (organochlorines (OCs), organophosphates (OPs), carbamates, and pyrethroids).

Organochlorines are organic compounds with five or more chlorine atoms. They were the first synthetic organic pesticides to be used in agriculture and in public health; they were widely used as insecticides for the control of insects.

4.1.1 Organophosphates

Organophosphates are phosphate acid esters or thiophosphoric acid esters their original compounds were highly toxic to mammals. Organophosphates are the general name for organic derivatives of phosphorus. They are the most commonly used insecticides in the world because their unstable chemical structure leads to rapid hydrolysis and little long-term accumulation in the environment [48]. Organophosphate manufactured since then is less toxic to mammals but toxic to target organism, such as insects. Some examples of organophosphate pesticides are malathion, parathion, diazinon and dichlorvos, tribufos (DEF), vamidothion, thiometon, and oxydemeton methyl.

4.1.2 Carbamates

Carbamates are a class of insecticides structurally and mechanistically similar to organophosphate (OP) insecticides. Carbamates are N-methyl Carbamates derived from a carbamic acid and cause carbamylation of acetylcholinesterase at neuronal synapses and neuromuscular junctions, some of the carbamates are aldicarb, carbaryl, oxamyl and terbucarb [49].

4.1.3 Pyrethroids

Pyrethroids are among the most frequently used pesticides and account for more than one-third of the insecticides currently marketed in the world [50]. Pyrethroids are known for their fast knocking down effect against insect pests, low mammalian toxicity, and facile biodegradation [51]. The synthetic pyrethroids with the basic structure of cyclopropane carboxylic ester are called type I pyrethroids. The pyrethroid insecticides were enhanced further by the addition of a cyano group at the benzylic carbon to give α-cyano are called type II, e.g., cyphenothrin and cypermethrin, tefluthrin.

The other major practice to pest management is biopesticides that are also a type of integrated pest management (IPM): biopesticides generally perform particularly well in IPM systems. With their lower toxicity profile, they are compatible with the use of classical biological control agents. Because they often are most effective at low pest pressures, they are well suited to be used in combination with scouting and monitoring activities, which detect pest problems before they are out of control. As well, IPM programs which include the rotation of biopesticides with conventional chemical pesticides can reduce reliance on single chemistries and delay the development of resistance within pest populations [52].

4.2 Biopesticides fall into three major classes

1. Biochemicals are naturally occurring substances (semichemical, plant extracts, minerals, PGRs, and organic acids) that control pests by nontoxic mechanism. Biochemical pesticides include substances that interfere with mating, such as insect sex pheromones. For example, neem (*Azadirachta indica*), garlic (*Allium sativum*), eucalyptus (*Eucalyptus globulus*), turmeric (*Curcuma longa*), tobacco (*Nicotiana tabacum*), and ginger (*Zingiber officinale*) have been successfully used for the management of several plant diseases [53].

2. Microbial pesticides consist of microorganism (e.g., a bacterium, fungus, protozoan). Microbial pesticides can control many different kinds of pests. The most widely used microbial pesticides are subspecies and strain of *Bacillus thuringiensis* (Bt). Almost 90% of the microbial biopesticides currently available on the market are derived from only onetime pathogenic bacterium, *Bacillus thuringiensis* (Bt) [54].

3. Plant-incorporated protectants (PIPs) are pesticide substances that plants produce from genetic material which has been added to the plant. A scientist can take the gene for the Bt pesticide protein and introduce the gene into the plant's own genetic material. Then, the plant manufactures the protein that destroys the pest. Pest resistance is one of the most widely targeted traits in plant genetic modification [55]; in Arizona (2010) cotton genetically modified by inserting Bt toxin from (*Bacillus thuringiensis*) to fight the pink bollworm moth (*Pectinophora gossypiella*) combined the release of sterile moth with growing genetically modified Bt cotton. This combined strategy reduced the need for insecticide spray and reduced pink bollworm abundance by 99%, with no increase in resistance to Bt cotton [55].

Tri-State Greenhouse IPM Workshop [6] reported that, recently, new substance has been reported as promising compounds for use as biopesticides. Extract of the Saponaria officinalis root and the nanoparticles showed a very good acaricidal efficacy [57], the fungus strains of *Talaromyces flavus* SAY-Y-94-01 [58], the fungus *Trichoderma harzianum*, fermentation products of the bacterium *Lactobacillus casei* strain LPT-111 [6], stilbenes accumulated in grape canes [50], and olive mill wastes [51].

In recent years, a new technology began to take place in IPM program; it could contribute to the development of less toxic biopesticides with favorable safety profiles and increased stability of the active agent, enhanced activity on target pest, and increased adoption by the end-users [43, 59]. Nanotechnology will contribute to making agriculture eco-friendlier and more profitable by reducing the usage of crop protection chemicals. Intelligent delivery of fertilizers, pesticides, and growth regulators, including nanosensors for real-time monitoring of soil conditions, crop growth, and pest and disease attack, is made possible through the development of nanodevices and products [60].

4.3 Nanopesticides

New practice that is increasing in the field of agriculture took the place for the potential to reduce the impact of other agrochemicals on human health and in the environment; the application sector of these nanomaterials is "nanopesticides" (ISO TC 299 "International standards for nanotechnology"). The European Commission (2011) based on the JRC report [61] defines that nanomaterials means "a natural, incidental or manufactured material containing particles, in an unbound state or as an aggregate or as an agglomerate and where, for 50% or more of the particles in the number size distribution, one or more external dimensions is in the size range 1 nm to 100 nm. In specific cases, and where warranted by concerns for the environment, health, safety or competitiveness, the number size distribution threshold of 50% may be replaced by a threshold between 1 and 50%."

ISO TC229 has published six technical specifications on nanotechnology terminology so far, namely:

ISO/TS 27687: 2008 Nanoobjects—Nanoparticle, nanofiber, and nanoplate.

ISO/TS 80004-1: 2010 Core terms.

ISO/TS 80004-2: 2010 Carbon nanoobjects.

ISO/TS 8004-3: 2011 Nanostructured materials.

ISO/TS 8004-4: 2011 Nano-/biointerface.

ISO/TS 8004-7: 2011 Diagnostics and therapeutics for healthcare [62].

Nanoscale material helps to reduce degradation of pesticide and fungicide and increase the effectiveness of application with reduced amount [45].

5. Toxicology of pesticides

Widespread use of pesticides is a significant source of air, water, and soil pollution causing risk to human health as a result of misuse or accident as well as leaving lasting harmful chemicals in the environment [63]. Also, effects of agricultural pesticides on nontarget organisms continue to become a major problem. Indiscriminate and injudicious use of chemical pesticide in agriculture has resulted in several associated adverse effects as environmental pollution, ecological imbalance, and pesticide residues in food, fruit, vegetable, fodder, soil, and water pest resurgence [64].

The WHO [46] grouped pesticides according to the potential risks to humans caused by accidental contact to human being to five classes:

Class Ia. Extremely dangerous parathion, dieldrin.

Class Ib. Highly dangerous eldrin, dichlorvos.

Class II. Moderately hazardous DDT, chlordane.

Class III. Slightly hazardous malathion.

Class IV. Products unlikely to present acute hazard in normal use.

The majority of pesticides are not specifically targeting the pest, during the application nontarget plants and animals are also affected, only about 0.1% pesticides reach the target organism, and the remaining applied pesticides contaminate the surrounding environment [63, 65].

Toxicity can be either acute or chronic:

a. Acute toxicity is the capability of a substance to cause harmful effects rapidly following exposure (few hours to a day).

b. Chronic toxicity is the capability of a substance to cause undesirable health effects resulting from long-term exposure [66].

5.1 Effects on humans

The main purpose of IPM is to reduce the effects of pests on crop product and help meet the increasing demand of larger population around the world. Although the application of pesticides achieves the goals of its usage, but at the same time, side effects also appear because of this practice.

Application of pesticides is a major threat to human health. It can taint food, water, soil, and air, causing headaches, drowsiness, fertility issues, and life-threatening illness; hundreds of thousands of known deaths occur each year due to pesticide poisoning [67]. Pesticide use has contributed toward improving agricultural production, in both yield and quality. Pesticides are also widely used in a variety of other settings, some of which most of the general public are not aware of [68]. It is evident that workers who are involved in mixing, loading, transport, and application of pesticide are at the highest risk of pesticide injury [69]. Pesticides can enter into the human body in three ways: (a) through the mouth (oral administration), (b) by adsorption through the skin or eyes (dermal adsorption), and (c) by breathing (inhalation) [70].

Also, atmospheric pesticides can cause hazards to humans. Atmospheric movement may cause transportation of pesticides from application sites to sensitive areas and accumulation of pesticides in the environment [71].

Risk related to pesticide poisoning can be defined as the extent of getting exposed to pesticide with a certain degree of toxicity. These can be expressed as Risk = Toxicity × Exposure [70].

5.1.1 Organochlorine pesticides (OCPs)

Organochlorine pesticides (OCPs) show multiple effects on the major physiological systems of the body including nervous, circulatory, and reproductive system and, also at some critical growth periods, may generate severe health disturbances [72].

Organophosphorus compounds are commonly used as insecticides. Organophosphate inhibits AChE, an enzyme located in the postsynaptic membrane that degrades AChE into choline and acetic acid [73]. The enzyme is classified as a B esterase whose function is the hydrolysis of acetylcholine which is a major neurotransmitter in the peripheral and central nervous system. The inhibition disturbs the capability of the enzyme to bind to its normal substrate with the subsequent accumulation of AChE at the nerve ending [74, 75]. The systematic investigation of the relationship between chemical structure and inhibition of AChE is the single most important feature required in an organophosphate for anticholinesterase activity and chemical reactivity; it has revealed a direct relationship between anticholinesterase activity and reactivity of the phosphorus atom [76]. Also, OPs can cause a type of toxicity called organophosphate-induced delayed polyneuropathy (OPIDP). It is characterized by deterioration of the long axons in the central and peripheral nervous system and ends with ataxia and paralysis which appear about 2–3 weeks after exposure [74]. Another side effect of Ops is oxidative stress and apoptosis. The damage is generated by the imbalance between reactive oxygen species (ROS) production and elimination [77]. Another side effect of Ops on human health is the disruption of estrogen function by acting as a ligand for receptor, converting other steroids to active estrogen or increasing the expression of estrogen-responsive genes [78]. Other Ops are capable of interfering with the endocrine function by inhibiting the binding of thyroid hormones to their corresponding receptors [78, 79]. Reiss [80] found out that the critical exposure period to PO insecticides for human neurological development is, by definition, the only relevant exposure for birth outcomes. Naksen et al. [81] reported that the birth outcomes as a result of OP exposure suggest a decrease in birth weight and head circumference in newborns born from mothers with low PONI activity.

5.1.2 Carbamates

Carbamates are hepatically metabolized via hydrolysis, hydroxylation, and conjugation, and 90% is renally excreted in a matter of days. The data of carbamates on central nervous system (CNS) and cerebrospinal fluid penetration, adults tend to have less CNS toxicity, whereas, in pediatric exposures, CNS depression is often a predominant symptom. Carbamates do not undergo aging that occurs during the phosphorylation of organophosphate to acetylcholinesterase and the carbamate-cholinesterase hydrolysis spontaneously within hours [82]. Fukuto [76] found out that insecticide carbamate causes AChE inhibition by identical mechanism to that of Ops. Unlike Ops poisoning, carbamate poisoning tends to be of shorter duration because the inhibition of nervous tissue acetylcholinesterase is reversible, and carbamates are more rapidly metabolism [83]. Forde [84] studied the effects of pregnant women exposure to carbamate; the results appear to show that carbamate when associated with other pesticides is typically used as OPs and pyrethroids. The result obtained is often related to OPs and pyrethroids. Carbamates are usually considered to be of limited acute toxicity.

5.1.3 Pyrethroids

The toxic effects of pyrethroids include neurotoxicity, skin contact, and respiratory and reproductive system toxicities [77]. Type I pyrethroid typical effects include rapid onset of aggressive behavior and increased sensitivity to external stimuli, followed by fine tremor, prostration with coarse whole-body tremor, elevated body temperature, coma, and death [85]. Type II pyrethroid effects are typically characterized by pawing and burrowing behaviors, followed by profuse salivation, increased startled response, abnormal hindlimb movement, and coarse whole-body tremors that progress to sinuous writhing. Clonic seizures may be observed prior to death; the term CS-syndrome (from choreoathetosis and salivation) has been applied to type II responses [85]. Gliga et al. [86] assessed the effects of three major herbicides, three insecticides, and three fungicides on three human cell lines (HepG2, HEK294, and JEG3); they found that fungicides were the most toxic from concentration 300–600 times lower than agricultural dilution, followed by herbicides and then insecticides, with very similar profiles in all cell types. LEG3 was the most sensitive cell line.

Nonoccupational low-dose exposure of any pesticides causes chronic disease in humans and can be considered as a silent killer; almost every crop faces a number of applications of different pesticides which results into multi-residue exposure of these pesticides that could be more in causing toxicity effects [87].

5.2 Effects on plants

Plants are the primary source of food for humans through crop production; crop safety and crop productivity are of paramount importance to ensure providing sufficient and healthy food for peoples.

Plants were the main reasons for pesticide application and practices, but in the early days of chemical pesticide applications, there were little concern about the side effects of this practice until illness started to appear on farmers and farm workers who are directly exposed to pesticides and using crop products that are treated with chemical pesticides. These effects alarmed governments, agriculture intuitions, and scientists around the world to pay a greater attention to these chemical pesticides used for crop protections.

Anonymous [88] found that absorption is the take-in of chemical substance into plants or microorganism. Most chemical pesticides break down once they are absorbed; pesticide residues may separate into simpler substances or remain inside the plant or animals and be released into the environment when the animal dies or plant decays.

A result of the study of Nishisaka et al. [89] to assess the genotoxicity effect of nanoparticles containing the paraquat herbicide indicated less chromosome damage than conventional paraquat herbicide. Saha and Gupta [90] found out that metallic nanoparticle, e.g., Ag NPs, causes significant toxic effects in animal cell culture and animal models; the impact of (Ag NPs) on plant species is related to oxidative stress-related gene expression, genotoxicity, seed germination, and root elongation; and genotoxic city studies revealed different types of chromosomal abnormalities and DNA damages which ultimately lead to cell death and disintegration of plant cell exposed to different coated and uncoated Ag NPs present in the environment. Toxicity of nanoparticles depends upon various factors like plant species, size, and concentration of nanoparticles in different stages of crops; it also depends on their composition and size. Small-sized nanoparticles are more reactive and toxic than the large-sized ones and affect the respiration or photosynthesis process [91]. For example, AL_2O_3 NPs showed phytotoxicity only on corn, reducing the root elongation by 35%. All improved root growth of grape and radish and inhibited root elongation of ryegrass and lettuce but had no effect on cucumber [92]. Boonyanitipong et al. [93] assess the effect of ZnO NPs on rice plant; the result shows adverse effect on rice from 100 mg/L and fully inhabit root growth and biomass at 500–1000 mg/L concentration. In a study of the effect of TiO_2 nanoparticle on aquatic life, the result raveled that TiO_2 reduced the light to entrap the algal cell and thus reduce the growth [94].

5.3 Effects on environment

The Environment Protection Act (EPA) (1986) defined the term environment under Section 2(a) of "to include water, air, land and inter-relationship between water, air, land and human being, other living creatures, plants, microorganisms and property." The definition includes complex relationship between environment parts; these parts must be in balance to insure healthy and accurate relationship.

Any disturbance in these relationships may lead to undesirable result. One of the most effected factors that play great roll in this disturbance is the application of different types of pesticides.

The potential for misapplication and accidental exposure is great [64]. It is found that only a very small part of the total amount of pesticides applied for weed and pest control (<0.1%) actually reaches the sites of action [95]. The runoff from agriculture and urban land, and rain precipitation and dry disposition from the atmosphere, can transport pesticides to streams and groundwater [96]. Mahmood et al. [97] reported that excessive use of pesticides may lead to the destruction of biodiversity. Birds, aquatic organism, and animals are under the threat of harmful pesticides. The soil is an important part of the environment and plays an effective role in other parts. The application of pesticides results into two ways: positive way by destroying the specific target and negative way by transferring to another non-specific target. In a study of Cessna et al. [98], they found that pesticides enter to the atmosphere by application drift, post-application vapor losses, or wind erosion of pesticide-treated soil; also, their photodegradation may be transported in long distances before the removal processes of atmospheric wet and dry deposition return them to the earth surface. Pesticides that were detected in the atmosphere are (I) organochlorine insecticides (resistant to environmental degradation), (II) organophosphate insecticides (not long lived in the environment), (III) atrazine herbicides (heavily used herbicides, persistent in the environment), (IV) acetanilide herbicides (used heavily, but not as persistent as atrazine) [71]. Mobility may result in redistribution within the application site and sometimes off-site. After application, a pesticide may (I) attach to soil particles, vegetation, or other surfaces and remain near the site; (II) attach to soil particles and move with eroded soil in runoff or wind; (III) dissolve in water and be absorbed by plants, overflow, or leach; (IV) pass off in vapor or erode from foliage or soil with wind and become airborne [99]. Also, the mobility of pesticides can be affected by several factors of pesticide sorption, water solubility, vapor pressure, and other environmental and site characteristics including weather, topography, canopy, ground cover, soil organic matter, texture, and structure [99]. The persistence of pesticide is expressed in terms of half-life that can help estimate whether or not a pesticide tends to build up in the environment. Pesticide half-lives are classified into three groups: low (less than 16-day half-life), moderate (16–59 days), and high (over 60 days). Pesticides with shorter half-lives tend to build up less and less likely to persist in the environment, while pesticides with longer half-lives are more likely to build up after repeated application. Higher persistence increases the risk of contamination of nearby surface water, groundwater, plants, and animals. Anonymous [88] reported that some pesticides stay in the soil long enough to be absorbed by plants grown in the field years later. The behavior of pesticides in soil is governed by a variety of complex dynamic physical, chemical, and biological processes, including sorption-desorption, volatilization, chemical and biological degradation, uptake by plants, runoff, and leaching [100, 101].

Biopesticides have benefits and limitation effects on the environment, human life, or agricultural product. They are highly effective in managing pests and diseases, without creating negative impacts on the environment, and their active and inert ingredients are generally recognized as safe. Besides the microbial content, carrier media for formulating biopesticide were consisted of several organic materials, such as animal broth, organic materials, or organic waste product. The media is a biodegradable material. In addition, biopesticides support stability and sustainability of agroecosystem because they did not affect negatively on the environment [102].

The nanoagrochemical is crucial to modern agriculture, and due to their direct and intentional application in the environment, nanoagrochemical may be regarded as particularly critical in terms of possible environmental impact, as they would represent the only intentional diffuse source of engineered nanoparticles in the environment [103]. There is harmful chemical reaction and contamination by nanoparticles to soil ecosystem and change in soil structure due to their large surface area and Brownian motion [45]. Kah et al. [104] assess the environmental fate of nanopesticides; the result suggests that the photodegradation and sorption behavior of clothianidin may have a greater impact on the environmental fate of pesticide AI than commercial formulations. AI clothianidin was rapidly released from the nanocarrier systems and that the durability of three nanoformulations would be short in water as well as in soil. Nanoparticles can easily be released in the water body or air, and uptake by living organisms creates toxic effect for humans and animals [43]. Bai et al. [105] found that CU nanoparticles caused damage in the central nervous system. Gliga et al. [86] study the effects of Ag nanoparticles, and the results show that Ag particles of size 10 nm were found more cytotoxic than other sizes.

6. The future

Pesticides are essential to improve the production of crops. The quantity of pesticides will continue to increase as long as the use of pesticides increases. Despite the tremendous benefits of pesticides for human beings especially in agriculture fields, side effects and undesirable results of pest managements such as pesticide residue crop products that are used in feed lead to several human illnesses in soil and water, microflora in soil, and ecosystem in general. Not until the year 1962 when biologist Carson published her book *Silent Spring* when dichlorodiphenyltrichloroethane (DDT) was at its high production 82 million Kg/year in the United States, it was initially used with great effect to combat malaria, typhus, and the other insect-borne human diseases among both military and civilian populations. The book inspired public concern about the toxicity in wildlife, contamination, and the increasing pest resistance. Control of regulated or quarantined pests is typically done through prevention of entry to a country or an area, eradication and containment, and use of tools such as biological control, pesticides and biopesticides, plant resistance, cultural methods, and natural enemy encouragement. In 2016 a review suggested that classical biological control has provided and should continue to provide many positive outcomes for dealing with damaging invasive alien insect pests [106].

Genetically modified (GM) food is a new type of potentially safer food without the use of pesticides; crops producing pesticides substance from genetic material that has been added to the plant. To insure safety, the EFSA Panel on Genetically Modified Organisms (GMO) require scientific risk assessment on the possible risk they might present for humans, animal health, and the environment before being authorized for market placement [107]. Also, the OECD Working Group for the Safety of Novel Foods and Feeds (WG-SNFF) addresses aspects of the safety assessment of food and feeds derived from genetically engineered crops. Their primary aim is promoting the use of consistent methods and data elements used in the risk/safety assessments among countries. The approach is to compare transgenic crops and derived products with similar conventional ones that are already known and considered safe for use, based on recognized practices, harmonized methods, and data sharing facilitated through the WG-SNFF [108].

Maximizing pesticide efficiency requires the use of radiolabeled pesticides to study pesticide metabolism, fate, residues, and formulation [109]. An increasing number of countries started to develop control strategies for the use of pesticides. The Danish National Action Plans on pesticide (2017–2021) strategy were (1) authorization of pesticides, (2) targeted inspection efforts, (3) collection of knowledge via the pesticide research program, and (4) information, advice, and guidance.

The Report of the OECD Workshop on Sustainable Pest Management in Practice: Anticipating and Adapting to Changes in the Pesticides Regulatory Landscape status and subsequent availability of agricultural pesticide products are necessary for sustainable pest management, including the use of registered agricultural pesticides. In general, the consequences of regulatory decision and the entailing process of adaptation of the agricultural production are not widely considered within the registration process. Regulators, pesticide manufactures, and pesticide users in OECD member countries have had to adapt their practices to ensure that sustainable and effective pest management options remain possible. These changes reduce risk to human health and the environment while promoting sustainable agriculture [110]. The Secretariat 2017 [111] in their 34 sessions includes recommendation that:

1. The international community must work on the development of a comprehensive, binding treaty to regulate hazardous pesticides throughout their life cycle. It should cover standardization among countries, policies to reduce pesticide use worldwide, and development of a framework for the banning and phasing-out of highly hazardous pesticides, as well as strict liability on pesticide producers.

2. Development of comprehensive national action plans to support alternatives to hazardous pesticides along with binding and measurable reduction targets and time frames.

While pesticides proved effective in mitigation of harmful bugs, the risk associated with their use has exceeded their beneficial effects. Nonselective pesticides can harm nontarget plants and animals along with the targeted ones; also with repeated use, some pests develop genetic resistance to pesticides [97].

To control the use of pesticides and reduce their effects, registration is an important aspect of pesticide management to ensure that the pesticide products released in the market are authorized and used only for their planned purpose. It will also enable authorities to implement controls for the price, packaging, labeling, safety, and advertisement of pesticides to ascertain protection of the user's interests [112].

To reduce pesticide impact on the environment, minimize contamination, and ensure the safety of human sources of food and water (surface and groundwater), users should be:

1. Practicing IPM

2. Using only pesticides that are labeled for their intended crop and pest

3. Considering application site characteristics and location of wells, ponds, and other water bodies

4. Maintaining application equipment, measuring, and calibrating accurately

5. Preventing back siphoning and spills, leaving buffer zone around sensitive areas, and reducing off-target drift

6. Considering the impact of weather/irrigation

7. Storing pesticides and disposing of wastes securely and safely [98–113]

Biopesticides have attracted attention in pest management in recent decades and have long promoted as prospective alternative to synthetic pesticides [68]. Although biopesticide use at a global scale is increasing by almost 10% every year [114], the global market must increase further in the future if these pesticides are to play a visible role in substituting for chemical pesticides and reducing the current overreliance on them [115]. It is expected that biopesticides will equalize with synthetics in terms of market size, between the late 2040s and the early 2050s [116]. Also, Soesanto [101] The conclusion of biopesticides was that biopesticides are the best way to control plant pathogens because of their beneficial effects; though there are still many limitations to be reduced, biopesticides supported stability and sustainability of agroecosystem because they did affect negatively on the environment.

Nanotechnology is the new type of IPM providing a promising future in the direction of formulation that can be used to improve the stability and effectiveness of natural product [117, 118]; it provides controlled release of the molecules at the site of action, can minimize potential toxic effects on nontarget organisms, and can prevent degradation of the active agent by microorganisms [118, 119]. Nanotechnology that includes nanopesticides seems to have a promising future in IPM. The potential toxicity of these nanoparticles is not standardized and not well understood yet explored by international and national safety regulators [60, 120–122]. Athanassious et al. [60] report that safer nanopesticides as alternative methods and practice should take the following into consideration: (a) The process of nanomaterial synthesis may cause changes in dimensions and shape; therefore, risk assessment studies are essential before the use of such materials. (b) Specific guidelines explain how to use these formulations on nanomaterials. (c) The toxic nature of these compounds to plants and insects needs to be analyzed.

(d) Working on nanopesticide formulation before they become more popular in pest management by combining analytical techniques that can detect, characterize, and quantify the active ingredient and adjuvants emanating from the formulation.

Emerging pesticide nanoformulations are not only increasingly complex and biologically active but may also exhibit a potential change in the physicochemical properties and/or biological effects at a size range that is larger than the nanoscale (>100 nm) [30].

-Tool and techniques to characterize properties (particle shape, size range, surface properties) of complex formulations of nanopesticides are lacking. Nanopesticides are more complex products by design and therefore pose greater challenges to analysts [30].

7. Conclusion

Continuous growth in population around the world leads to increase the demand for higher crop production. The quality and quantity of crops provided to people must be satisfactory, which can be achieved by using specific methods to control pests that play a great role in crop losses and poor product. The main method used for this purpose is synthetic pesticides, with other methods: biopesticides and nanopesticides. Despite the harmful side effects especially of synthetic pesticide compared to the other methods with less harmful effects on humans, plants, and the environment, still the synthetic pesticides play an important part of IPM. This requires intensive work of scientists, institutions of agriculture around the world, environment studies to assess and evaluate the side effects of these different methods, and provide good training for safer application of pesticides and also continues studies for every new chemical production and methods used in the agriculture field, to decrease and minimize the harmful effects on humans, animals, plants, nontarget organisms, and the environment, including aquatic environment.

Conflict of interest

The author declares no conflict of interest.

Author details

Mona Saud AL-Ahmadi
Department of Biology, Science College, Imam Abdulrahman Bin Faisal University, Dammam, Saudi Arabia

*Address all correspondence to: dr.alahmdi2009@yahoo.com

References

[1] UN DESA. World population projected to reach 9.7 billion by 2050. United Nations. Department of Economic and Social Affairs. 2015. p.3Co2-earth. Available from: https://www.co2.earth/

[2] Al-Sadi AM. Variation in resistance to spot blotch and the aggressiveness of Bipolaris sorokiniana on barley and wheat cultivars. Journal of Plant Pathology. 2016;**98**:97-103. DOI: 10.4454/jpp. v98il.029

[3] Balique F, Lecoq H, Raoult DE, Olson P. Can plant viruses cross the kingdom border and be pathogenic to humans? Viruses. 2015;7:2074-2098. DOI: 10.3390/v70422074

[4] Al-Sadi AM. Impact of plant diseases on human health. International Journal of Nutrition, Pharmacology, Neurological Diseases (Serial Online). 2017;**7**:21-22

[5] Bolognesi C. Genotoxicity of pesticides: A review of human biomonitoring studies. Mutation Research. 2003;**543**(3):251-272

[6] Tri-State Greenhouse IPM Workshop. Biopesticides and their Modes of Action. 2018

[7] Oerke EC, Dehne HW, Schonbeck F, Weber A. Crop Production and Protection: Estimated Losses in Major Food and Cash Crops. Amsterdam, Netherlands: Elsevier. Science Ltd; 1994. p. 808

[8] Damalas A. Understanding benefits and risk of pesticide use. Scientific Research and Essay. 2009;**4**(10):945-949

[9] Hendery S. Using Integrated Pest Management to Reduce Pesticides and Increase Food Safety. Integrated Pest Management Innovation Lab; 2018. Available from: https://www.agrilinks.Org/users/ipm-innovation-lab# profile-main

[10] Maksymiv I. Pesticides: Benefits and hazards. Journal of Vasyl Stefanyk. Precarpathian National University. 2015;**2**(1):70-76. UDC 632.95.02+57.044. Doi: 10.15330/ jpnu. 2.1.70-76

[11] Miller GT. Sustaining the Earth. 6th ed. Pacific Grove: Thompson Learning Inc.; 2004

[12] Gonzalez-Coloma A. Biopesticides: Biotechnology and natural products chemistry. Revista de ProtecciÓn Vegetal. Open Journal. 2015;**30**:30

[13] Sharma S, Malik P. Biopesticides: Type and applications. International Journal of Advanced in Pharmacy, Biology and Chemistry. 2012;**4**:515-508

[14] Chhipa H. Nano-pesticides: Current status and future possibilities. Agricultural Research & Technology. Open Access Journal. 2017:**5**(1):1-4. DOI: 10.19080/ ARTOAJ.2017.05.555651

[15] Kumar S, Singh A. Biopesticides: Present status and the future prospects. Journal of Fertilizer & Pesticides. 2015;**6**(2):100-129. DOI: 10.4172/2471-2728.1000e129

[16] Gonzales JOW, Stefanazzi N, Murray AP, Ferrero AA, BB F. Novel Nano insecticides based on essential oils to control the German cockroach. Journal of Pest Science. 2015;**88**:393-404

[17] Chhipa H, Joshi P. Nonfertilizer, nano-pesticides and nano sensors in agriculture. In: Ranjan S, Dasgupta N, Lichtfouse E, et al., editors. Nanos eScience in Food and Fuel. New Delhi, India: TERI; 2015

[18] Venugopal NV, Sainadh NV. Novel polymeric nano formulation of

mancozeb—An eco-friendly nanomaterial. International Journal of Nanoscience. 2016;**15**(4):1650016

[19] Small M. Plant Pathology. Master Gardener. Colorado State University. 2017. CMG GardenNote#331

[20] Tang PS, Krupa SV. Crops loss assessment. In: Proceeding of E.C. Stakman Commemorative Symposium. Minnesota Environmental Quality Board Power Plant Siting Program. 1980

[21] Teng PS. Crops Losses Assessment and Pest Management. St. Paul, MN: APS Press; 1987

[22] Oerke EC. Crop losses to pests. Journal of Agricultural Science. 2006;**144**:31-43. DOI: 10.1017/S0021859605005708

[23] Oerke EC, Dehne HW. Safeguarding production—Losses in major crops and the role of crop protection. Crop Protection. 2004;**23**:275-285

[24] Strange RN, Scott PR. Plant disease: A threat to global food security. Annual Review of Phytopathology. 2005;**43**: 83-116. DOI: 10.1146/annurev.Phyto.43.113004.133839

[25] FAO. Intergovernmental Technical Panel on Soils of the Global Soil Partnership. Pesticides and Plant protection. Food and Agriculture Organization of The United Nations. Rome, Italy; 2017

[26] Bajwa WI, Kogan M. Compendium of IPM Definitions (CID)—What is IPM and how is it defined in the Worldwide Literature? IPPC Publication No. 998. Oregon State University. Corvallis, OR, USA: Integrated Plant Protection Center (IPPC); 2002

[27] Flint ML, Bosch R. Introduction to Integrated Pest Management. New York: Plenum Press; 1981. p. 240

[28] Ghandler D, Alastair S, Bailey G, Tatchell M, Davidson G, Greaves J, et al. The development, regulation and use of biopesticides for integrated pest management. Philosophical Transactions of the Royal Society B. 2011;**366**:1987-1998. DOI: 10.1098/rstb.2010.0390

[29] US EPA. Reregistration Eligibility Decision for Aldicarb. 2005. Available from: http://www.epa.gov/pesticides/reregistration/REDs/aldicarb_red.pdf - Accessed 12/14/09

[30] Kookana RS, Ashauer R, Beulke S, Chaudhry Q, Cornelis G, Fernanes Gan J, Kah M, Ranville J, Van den Brink PJ. Nano-pesticides: Guiding principles for regulatory evaluation of environmental risks. Journal of Agricultural and Food Chemistry. 2014;**62**(19):4227-4240. DOI: 10.1021/jf500232f

[31] Zacharia JT. Identity, Physical and Chemical Properties of Pesticides, Pesticides in the Modern World- Trends in Pesticides Analysis. In Tech; 2011. Available from: http://www.intechopen.com/books/pesticides-in. DOI: 10.5772/17513

[32] FAO. Sustainable development and natural resources management. In: Twenty-Fifth Conference, Paper C 89/2-Sup. 2. Rome. 1989

[33] Garcia FP, Sandra Y, Ascencio C, John C, Oyarzun G, Hernadez AC, et al. Pesticides: Classification, uses and toxicity. Measures of exposure and genotoxic risks. Journal of Research in Environmental Science and Toxicology. 2012;**1**(11):279-293

[34] Grube A, Donaldson D, Kiely T, Wu L. Pesticides Industry Sales and Usage. 2006 and 2007 Market Estimates. Washington, DC: U.S. Environmental Protection Agency; 2011

[35] Seed Quest. Global Information Services for Seed Professionals. Market

Data and Statistics. 2016. Available from: http://www.seedquest.com

[36] Hellar H. Pesticides Residues in Sugarcane Plantations and Environ After Long-Term Use. Kilimanjaro Region, Tanzania: The Case of TPC Ltd; 2002

[37] Cooper J, Dobson H. The benefits of pesticides to mankind and the environment. Crop Protection. 2007;**26**(9):1337-1348

[38] Kumar R. Risking and wronging. Philosophy Public Affairs. 2015;**43**(1):27-51

[39] Marrone PG. The market and potential for biopesticides. In: Gross AD, Coats JR, Duke SO, Seiber JN, editors. Biopesticides: State of the Art and Future Opportunities. Washington, DC, USA: American Chemical Society; 2014. pp. 245-258

[40] The Annual General Meeting (AGM) of Bio-Works Technologies AB (pub), corporate identity number 556935-3559. 2018

[41] Sarwar M. Biopesticides: An effective and environmental friendly insect-pests inhibitor line of action. International Journal of Engineering and Advanced Research Technology. 2015;**1**(2):10-15

[42] Pérez-de-Luque A, Rubiales D. Nanotechnology for parasitic plant control. Pest Management Science. 2009;**65**:540-545

[43] Khot LR, Sankaran S, Maja JM, Ehsan R, Schuster E. Application of nanomaterials in agricultural production and crop protection: A review. Crop Protection. 2012;**35**:64-70

[44] Bergeson LL. Nano silver: US EPA's pesticide office considers how best to proceed? Environmental Quality Management. 2010;**19**(3):79-85

[45] Dubey A, Mailapalli D. Nonfertilizer, nano-pesticides, nano-sensors of pest nanotoxicity in agriculture. In: Lichtfouse E, editor. Sustainable Agriculture Reviews. Vol. 19. Switzerland: Springer International Publishing; 2016. DOI: 10.1007/978-3-319-26777-7_7. Available from: https://www.researchgate.net/publication/

[46] Drum C. Soil Chemistry of Pesticides. USA: PPG Industries, Inn.; 1980

[47] WHO. World Health Organization Study Group "Diet" Nutrition and the Prevention of Chronic Diseases. WHO Technical Report Ser. 799. 1990

[48] Soltaninejad K, Shadnia S. History of the Use and Epidemiology of Organophosphorus Poisoning. In: Balali-Mood M, Abdollahi M, editors. Basic and Clinical Toxicology 25 of Organophosphorus Compounds. London; Springer-Verlag; 2014:25-43. DOI 10.1007/978-1-4471-5625-3_2

[49] Dubois S, Fenwick N, Ryan EA, Baker L, Baker SE, Beausoleil NJ, et al. International consensus principles for ethical wildlife control. Conservation Biology. 2017;**31**(40):753-760

[50] Yadav RK, Chattopadhyay D. Differential soybean gene expression during early phase of infection with Mungbean yellow mosaic India virus. Molecular Biology Reports. 2014;**41**(8):5123-5134. DOI: 10.1007/s11033-014-3378-0. https://www.ncbi.nlm.nih.gov/pubmed/24752408

[51] El-Abbassi A, Saadaoui N, Kiai H, Raiti J, Hafidi A. Potential applications of olive mill wastewater as biopesticide for crops protection. Science of The Total Environment. 2017;**576**:10-21. DOI: 10.1016/j.scitotenv.2016.10.032

[52] Agriculture and Agri-Food Canada. Biopesticides. 2018. Available from:

http://www.agr.gc.ca/eng/science-and-innovation/agricultural-practices/biopesticides/?id=1531920003497

[53] Yadav RK, Chattopadhyay D. Differential soybean gene expression during early phase of infection with Mungbean yellow mosaic India virus. Molecular Biology Reports. 2014;**41**(8):5123-5134. DOI: 10.1007/s11033-014-3378-0

[54] Koul O. Microbial biopesticides: Opportunities and challenges. Perspectives in Agriculture, Veterinary Science, Nutrition and Natural Resources. 2011;**6**(056). DOI: 10.1079/PAVSNNR20116056

[55] Carrière Y, Ellers-Kirk C, Biggs R, Degain B, Holley D, Yafuso C, et al. Effects of cotton cultivar on fitness costs associated with resistance of pink bollworm (Lepidoptera: Gelechiidae) to Bt cotton. Journal of Economic Entomology. 2005;**98**(3):947-954

[56] Damalas D. A brief reflection on the Mediterranean fisheries: Bad news, good news and no news. Journal of Fisheries Research. 2017;**1**(1):28-33

[57] Mensah AY, Mireku EA, Okwuonu V. Anti-inflammatory and anti-oxidant activities of *Secamone afzelii* (Rhoem) Ascleipiadaceae. Journal of Medical and Biomedical Sciences. 2014;**3**(1):23-30. DOI: 10.4314/jmbs.v3i1.4"10.4314

[58] Ishikawa K, Ito K, Inoue J, Semba K. Cell growth control by stable Rbg2/Gir2 complex formation under amino acid starvation. Genes to Cells. 2013;**18**(10):859-872

[59] Prasad S, Tyagi AK. Recent developments in delivery, bioavailability, absorption and metabolism of curcumin: The golden pigment from golden spice. Cancer Research and Treatment. 2014;**46**(1):2-18. DOI: 10.4143/crt.2014.46.1.2.Epub

[60] Athanassious CG, Kavallieratos NG, Benelli G, Losic DU, Rani P, Deneux N. Nanoparticles for pest control: Current status and future perspectives. Journal of Pest Science. 2018;**91**:1-15

[61] Lövestam G, Rauscher H, Roebben G, et al. Considerations on a definition of nanomaterial for regulatory purposes. The Joint Research Centre of the European Commission. 2010

[62] APVMA Report. Regulatory Considerations for Nano-pesticides and Veterinary Nanomedicines. Australian Pesticides and Veterinary Medicines Authority. 2014

[63] Darcn E, Darcn M. Health effects of agriculture pesticides. Biomedical Research. 2017;(special issue):13-17

[64] Parween T, Jan S, Mahoodu SZ, Tasneem F, Siddiqui ZH. Selective effect of pesticides on plant—A review. Critical Reviews in Food Science and Nutrition. 2016;**56**:160-179

[65] Bell JNB, Power SA, Jarraud N, Agrawal M. The effects of air pollution on urban ecosystems and agriculture. The International Journal of Sustainable Development and World Ecology. 2011;**18**(3):226-235. DOI: 10.1080/13504509.2011.570803

[66] Pesticide Toxicity and Hazard. British. Colombia. Ministry of Agriculture. 2017:1-9. https://www2.gov.bc.ca/assets/gov/farming-natural-resources-and-industry/agriculture-and-seafood/animal-and-crops/plant-health/pesticide-toxicity-hazard.pdf

[67] Hendery S. Using Integrated Pest Management to Reduce Pesticides and Increase Food Safety. US Government's Global Hunger and Food Security Initiative. 2018. Available from: https://www.agrilinks.org/user/ipm-innovation-lab#profil-main

[68] Damalas C. Understanding benefits and risks of pesticide use. Scientific Research and Essay. 2009;**4**(10):945-949

[69] Fenske RA, Day EW. Assessment of exposure for pesticides handlers in agricultural, residential and institutional environments. Occupational and Residential Exposure Assessment for Pesticides. 2005:11-43

[70] Schulze LD, Ogg C, Vitzthum EF. Signs and Symptoms of pesticide Poisoning. EC97-2505. 1997

[71] Toth SJ, Buhler WG. Environmental Effects of Pesticides. Department of Entomology and Horticultural Science: North Carolina State University; 2009

[72] Singh Z, Kaur J, Kaur R, Hundal SS. Toxic effects of organochlorine pesticides: A review. American Journal of BioScience. 2016;**4**(3-1):11-18

[73] Cocco P. Pesticides and human health. Oxford Research Encyclopedia of Environmental Science. USA: Oxford University Press; 2016. pp. 1-78. DOI: 10.1093/acrefore/9780199389414.013.82

[74] Jokanovic M. Biotransformation of organophosphorus compounds. Toxicology. 2001;**166**:139-160

[75] Remington SE. Cellular response to DNA damage after exposure to organophosphates in vitro (A thesis for Doctor of Philosophy). BSc (Hons) Medical Biochemistry (Birmingham University). MSc Toxicology (Birmingham University); 2010

[76] Fukuto TR. Mechanism of action of organophosphorus and carbamate insecticides. Enviro Health Perspective. 1990;**87**:245-254

[77] Lei Wang DD, Dou TY, Hou J, Feng L, et al. Assessment of the inhibitory effects of pyrethroids against human carboxylesterases. Toxicology and Applied Pharmacology. 2017;**321**:48-65

[78] Mostafalou S, Abdollahi M. Pesticides: An update of human exposure and toxicity. Archives of Toxicology. 2017;**91**:549-599

[79] Androutsopulos VP, Hemandez AF, Liesivuoni J, Tsatsakis AM. A mechanistic overview of health associated effects of low levels of organochlorine and organophosphorus pesticides. Toxicology. 2013;**307**:89-49

[80] Reiss R, Chang ET, Richardson RJ, Goodman M. A review of epidemiologic studies of low-level exposures to organophosphorus insecticides in non-occupational populations. Critical Reviews in Toxicology. 2015;**45**:531-641

[81] Naksen W, Prapamontol T, Mangklbruks A, Ryan PB, Riederer AM, et al. Associations of maternal organophosphate pesticide exposure and PON1 activity with birth outcomes in SAWASDEE birth cohort. Environmental Research. 2015;**142**:288-296

[82] Silberman J, Taylor A. Carbamate Toxicity. NCBI Bookshelf. Treasure Island (FL): State Pearls Publishing; 2018

[83] Fishel FM. Pesticide toxicity profile: Chlorinated hydrocarbon. Pesticides. Department, UF/IFAS. 2011:1-3. http://edis.ifas.ufl.edu

[84] Forde MS, Roberison L, Laouan Sidi EA, Cote S, Gaudreau E, et al. Evaluation of exposure to organophosphate, carbamate, phenoxy acid, and chlorophenol pesticides in pregnant women from 10 Caribbean countries. Environmental Science: Processes & Impacts. 2015;**17**:1661-1671

[85] U.S. Department of Health and Human Services. Toxicological Profile for Pyrethrins and Pyrethroids. Agency for Toxic Substances and disease

Registry. Division of Toxicology/ Information Branch; Atlanta, Georgia. 2003

[86] Gliga AR, Skoglund S, Wallinder IO, Fadeel B, Karlsson HL. Size-dependent cytotoxicity of silver Nano-particles in human lung cell: The role of cellular uptake, agglomeration and Ag release. Particle and Fibre Toxicology. 2014;**11**: 1-17. DOI: 10.1186/1743-8977-11-11

[87] Ganguly P, Sigh VK. Pesticide and its risk assessment. Toxicology. 2018;**2**(3):1-2

[88] Anonymous. Environmental Protection, Environmental Fate. 2009. Available from: http://www.al.gov.bc.ca/pesticides/c_2htm

[89] Nishisaka C, Grillo R, Sanches G, Fraceto L. **A**nalysis of the effects of pesticides and Nano-pesticides on the environment. In: 5th Congress of Brazilian Biotechnology Society (SBBIOTEC), Brazil. Oct 14, 2014

[90] Saha N, Gupta SD. A glimpse on silver nano-particles genotoxicity in higher plants. Global Journal of Nano-medicne. 2017;**2**(2):1-2

[91] Navarro E, Baun A, Behra R, Harmann BN, Filser J, Ai-Jun M, et al. Environmental behavior and ecotoxicity of engineered nanoparticles to algae, plants, and fungi. Ecotoxicology. 2008;**17**:372-386. DOI: 10.1007/s10646-008-0214-0

[92] Daohui L, Xing B. Phytotoxicity of nanoparticles: Inhibition of seed germination and root growth. Environmental Pollution. 2007;**150**: 243-250. DOI: 10.1016/jenvpol.2007.01.016

[93] Boonyanitipong P, Kositsup B, Kumar P, Baruah S, Dutta J. Toxicity of ZnO and TiO_2 nanoparticles on germination rice seed *Oryza sativa* L. International Journal of Bioscience, Biochemistry and Bioinformatics. 2011;**1**(4):282-285

[94] Sharma VK. Aggregation and toxicity of titanium dioxide nanoparticles in aquatic environment—A review. Journal of Environmental Science and Health. Part A, Toxic/Hazardous Substances & Environmental Engineering. 2009;**44**(14):1485-1495. DOI: 10.1080/10934520903263231

[95] Pimental D. Amounts of pesticides reaching the target pests: Environmental impacts and ethics. Journal of Agricultural and Environmental Ethics. 1995;**8**:17-29

[96] Geological Survey U.S. Geological Survey. 2006. Mineral Commodity Summaries. USGS. Science for changing world United States Governments Printing Office; 2006

[97] Mahmood I, Imadi SR, Shazadi K, Hakeem KR. Effects of Pesticides on Environment. 2015. DOI: 10.1007/978-3-319-27455-3_1. https://www.researchgate.net/publication/286042190

[98] Cessna AJ, Wolf TM, Stephenson GR, Brown RB. Pesticide movement to field margins: Routes. Impacts and mitigation. Field boundary habitats: Implications for weed. Insect and Disease Management. 2005;**1**:69-112

[99] Kerle EA, Jenkins JJ, Vogue PA. Understanding pesticide persistence and mobility for groundwater and surface water protection. Oregon State Univ Extension Service. EM8561-E. 2007

[100] Aris-Estévez M, López-Periago E, Martnez-Carballo E, Simal-Gándara J, Mejuto Juan C, Garcia-Rio L. The mobility and degradation of pesticides in soils and the pollution of groundwater resources. Agriculture Ecosystems & Environment. 2008;**123**:247-260

[101] Tiryaki1 O, Temur C. The fate of pesticide in the environment. Journal

of Biodiversity and Environmental Sciences. 2010;**4**(10):29-38

[102] Soesanto L. Biopesticides: Impact to the environment. In: The 5th International Seminar of Indonesian Society for Microbiology. 2012

[103] Kah M, Beulke S, Tiede K, Hofmann T. Nano-pesticides state of knowledge, environmental fate and exposure modelling. Critical Reviews in Environmental Science and Technology. 2013;**43**:1823-1867. DOI: 10.1080/10643389.2012671750

[104] Kah M, Walch H, Hofmann T. Environmental fate of nano-pesticides: Durability, sorption and photodegradation of nanoformulted clothianidin. Environmental Science: Nano. 2018;**5**:882-889

[105] Bai R, Zhang L, Liu Y, Li B, Wang P, Autrup H, et al. Integrated analytical techniques with high sensitivity for studying brain translocation and potential impairment induced by intranasally instilled copper nanoparticles. Toxicology Letters. 2014;**226**:70-80. DOI: 10.1016/j.toxlet.2014.01.041

[106] Peterson1 JA, Ode PJ, Oliveira-Hofman C, Harwood JD. Integration of Plant Defense Traits with Biological Control of Arthropod Pests: Challenges and Opportunities. Frontiers in Plant Science; 2016;**7**:1-23. DOI: 10.3389/fpls.2016.01794. www.frontiersin.org

[107] EFSA Panel on Genetically Modified Organisms (GMO). Guidance for risk assessment of food and feed from genetically modified plants. EFSA Journal. 2011;**9**(5):37. Article ID: 2150. DOI: 10.2903/j.efsa.2011.2150

[108] OECD. Internal Co-ordination Group for Biotechnology—ICGB Newsletter No. 33. 2018

[109] Lindquist DA. Pesticides: Chemicals for survival. IAEA Bulletin. 1981;**23**(3):37-39

[110] OECD. Chemical Safety and Biosafety News. No. 37. 2018

[111] The Secretariat. Report of the Special Rapporteur on right to food. Human Right council. 34 session. General Assembly. United Nations. 2017

[112] WHO. International code of conduct on the distribution and use of pesticides: Guidelines for the registration of pesticides. World Health Organization, Rome; 2010

[113] Reichenberger S, Bach M, Skitschak A, Frede HG. Mitigation strategies to reduce pesticide inputs into ground and surface water and their effective eness; A review. Science of the Total Environment. 2007;**384**(1-3):1-35

[114] Kumar S, Singh A. Biopesticides: Present status and the future prospects. Journal of Fertilizers & Pesticides. 2015;**6**:e129

[115] Czaja K, Goralczyk K, Strucinski P, Hernik A, Korcz W, Minorczyk M, et al. Biopesticides-towards increased consumer safety in the European Union. Pest Management Science. 2015;**71**:3-6

[116] Damalas CA, Koutroubas SD. Current Status and Recent Developments in Biopesticide Use. Agriculture. 2018;**8**(13):1-6. DOI: 10.3390/agriculture8010013

[117] Ghormade V, Deshpande MV, Paknikar KM. Perspectives for Nano-biotechnology enabled protection and nutrition of plant. Biotechnology Advances. 2011;**29**:792-803

[118] Perlatti B, de Souza Bergo PL, Fernandes da Silva MF, Fernandes JB, Forim MR. Polymeric nanoparticle-based insecticides: A controlled release

purpose for agrochemicals. In: Trdan S, editor. Insecticides-Development of Safer and More Effective Technologies. Rijeka, Croatia: In Tech; 2013. pp. 523-550

[119] Gogos A, Kuauer K, Bucheli TD. Nanomaterials in plant protection and fertilization: Current state, foreseen application, and research priorities. Journal of Agricultural and Food Chemistry. 2012;**60**:9781-9792

[120] Benelli G. Research in mosquito control: Current challenges for a brighter future. Parasitology Research. 2015;**114**:2801-2805

[121] Murugan K, Benelli G, Panneerselvam C, Subramaniam J, Jeyalalitha T, Dinesh D, et al. *Cymbopogon citratus-synthesized* gold nanoparticles boost the predation efficiency of copepod *Mesocyclops aspericornis* against malaria and dengue mosquitoes. Experimental Parasitology. 2015b;**153**:129-138

[122] Murugan K, Dinesh D, Madhiyazhagan P, Panneerselvam C, Subramaniam J, Suresh U, et al. Predation by Asian bullfrog tadpoles, *Hoplobatrachus tigerinus*, against the dengue vector Aedes aegypti in an aquatic environment treated with mosquitocidal nanoparticles. Parasitology Research. 2015;**114**:3601-3610

Chapter 5

Uses and Misuses of Agricultural Pesticides in Africa: Neglected Public Health Threats for Workers and Population

Pouokam Guy Bertrand

Abstract

Pesticides are use in agriculture for their capacity to reduce pest and protect foods. Since their introduction in Africa by colonial masters, the use of these chemicals is constantly growing. Herbicides and insecticides are the two dominant categories. Although they are used in small quantities by farmers who own small exploitation, the frequency of their use, as well as overuses and misuses, constitutes serious factors of exposure and health risks. Farm workers are more vulnerable to occupational effects from pesticide inhalation and skin contacts. Failure to wear protective equipment and observe good agricultural practices explained health symptoms that are frequently experienced: eye and skin irritation, nausea, vomiting, and headache. Population is subject to chronic health effects due to repeated dietary intake of pesticides. Most consumed staple foods on the continent (cereals, vegetables, and fruits) have been found to be contaminated by one or multiple residues of pesticides. The level of residues is often higher than regulatory limits. Organization of surveillance programs to monitor concentration of pesticide residues remains inexistent in most countries, same for toxicovigilance systems to documented poisoning cases. Current data underline the need to carry out pesticide health risk assessment in order to appreciate the threats they pose for public health.

Keywords: pesticide, exposure, poisoning, residues, misuses

1. Introduction

Synthetic pesticides have been used in Africa since more than eight decades. They were probably introduced in the continent by colonial masters. From a historical point of view, the Public Health Act of the British government legislation, to protect human beings and regulate the use of pesticides by farmers in Kenya, was enacted in 1921 [1]. From that period till date, African countries continue to import pesticides from more advanced economies, mainly from European countries and recently from China.

The FAO statistical (FAOSTAT) database estimated that Africa have imported in 2016 pesticides for a value of 1590160.326 USD [2]. However, since few years, pesticides importation from China is constantly growing.

According to the export statistics from China Customs, export volume of pesticides (under HS code 29 and 38) during January to November in 2015, African markets represented 13.9% of the total export of pesticides from China and concerned 44 countries. Top 10 countries by export value are Nigeria, South Africa, Ghana, Ivory Coast, Egypt, Kenya, Cameroon, Tanzania, Ethiopia, and Guinea. The amount of export value for these top 10 countries constitutes 85.9% of the total export value to Africa from China, as shown in **Figure 1** [3].

Thousands of pesticides are imported, but the top 10 of most imported formulation products and active ingredients can be seen in **Table 1**.

Top 10 of most imported pesticides are mostly herbicides, followed by next 10 others products formulations which are mainly insecticides (chlorpyrifos, dichlorvos, dimethoate, abamectin, and cypermethrin).

It should be noted that in almost all countries, before their importation, pesticide products have to be homologated and authorized. Information are also provided to users on the purposes to which the products should be used, the dosage, their toxicological classes, first aid action to be taken in case of exposure, and even antidote in case of swallowed. Despite these precautions, Pouokam et al. [4] have reported several misuses in many countries and in particular in Cameroon.

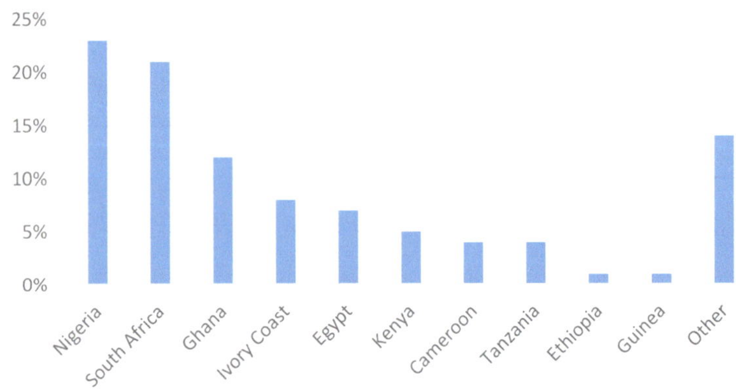

Figure 1.
Top 10 of pesticides export in Africa by export values (adapted from Agronews).

Formulation products	Export value (million USD)	Export volume (000 tons)
Glyphosate IPA 41% SL	93.82	50.82
Paraquat 200 g/L SL	62.53	25.97
Glyphosate 30%	20.07	10.44
Lambda-cyhalothrin 25 g/L EC	16.97	5.68
Mancozeb 80% WP	16.69	5.08
Glyphosate-monoammonium 75.7% SG	16.33	4.84
2,4-D-dimethylammonium 720 g/L SL	14.78	7.29
Mancozeb 64% + Metalaxyl 8% WP	11.61	1.80
Atrazine 80% WP	10.19	3. 13
Imidacloprid 20% SL	8. 13	1.91
Total	271.12	111.69

Table 1.
Top 10 of most imported formulation products and active ingredients (adapted from Agronews)

2. Uses and misuses of agricultural pesticides in Africa

Pesticide homologation procedures are established in all African countries. Procedures differ from one country to another, and the majority of pesticides formulations are homologated for uses in agriculture. In this chapter, we recall some key information concerning the number of pesticides formulations in distribution in the top 10 pesticides importers as indicated above.

In Cameroon, the official database from the Ministry of Agriculture counted up to 610 [4]. de Vos et al. reported 500 pesticide formulations used in South Africa Republic [5]. In 2011, the volume of imported pesticides in Ghana was 20,747 tons (9216 tons of insecticides, 8986 tons of herbicides, 2545 tons of fungicides) [6]. Ivory Coast and 13 other countries of the Permanent Interstate committee for Drought Control in the Sahel (CILLS), namely, Benin, Burkina Faso, Cap-Vert, Guinea, Gambie, Guinee-Bissau, Mali, Mauritania, Niger, Senegal, Tchad, and Togo, had adopted a common homologation procedures. The actual homologated list contains 438 pesticides formulations [7]. In 2016, Egypt consumed up to 10,600 metric tons of pesticides formulations [8]. Between July 2013 and June 2014, a total of 1182 different types of pesticides were registered in Tanzania, of which 11.2% were provisionally registered [9]. In 2017, 117 pesticide formulations were found to be homologated for used in Ethiopia, among which were 68 insecticides, 45 herbicides, and 4 fungicides [10].

Major uses of pesticides include agriculture, livestock development, and disease vectors control. Choice of pesticides depends on farmers' perception of its efficacy on pests, type and intensity of pests, crops growth stage and availability of pests, crops growth, and the availability of pesticide [11].

Supply channels of pesticides are both formal and informal and include:

- Authorized retail outlets of agricultural supply companies
- Government extension services
- Small-scale informal traders operating via local shops
- Itinerant peddlers visiting villages and weekly markets
- Bulk supplies from general markets in larger cities

A lot of misuses have also been reported in field farms because of (i) absence of clear instructions, (ii) illiteracy of farmers, (iii) lack of knowledge on risks from bad uses, (iv) uses of pesticides on crops for which the product was not homologated, and (v) difficulties to properly prepare the solution to be used, and poor respect of dosage [12].

In Benin Republic, previous studies, for example, revealed that farmers have been found to use insecticides registered for cotton protection on vegetables [12]. Moreover, Pouokam et al. observed the reuses of pesticides empty containers for drinking water and traditional wine. Other containers are washed and rinsed in rivers were populations fetch water for domestic uses [4].

Pesticide applications also vary with the type of crops and the type of equipment used, most farmers used knapsack and therefore carry the equipment on their back; others used atomizer, especially cocoa farmers. Because they did respect wind direction, they often received pesticides vapor in their eyes and bodies part.

A detail analysis of all these practices gives insight on potential risk factors of pesticide exposures both for workers and nearby populations.

3. Use of predictive pesticides exposure models for occupational and para-occupational scenarios

Practice of agriculture in Africa is dominated by small farmers and family farmers whose one characteristic is that they operate in the informal sector. They are not registered and most of the time remains unknown by local authorities. This absence of clear statute for agricultural workers makes it difficult to organized and regulate the sector. Another consequence is the poor identification of exposure and poisoning cases.

Naveen Kumar et al. defined pesticides exposures as the contact of the pesticides with a surface or an organism. For a human, it means getting pesticides in or on the body [13].

Occupational and para-occupational exposures to agricultural pesticides in Africa remain as key concern that are unfortunately poorly addressed. Occupational exposure occurred in work places and concerned workers, while the definition of para-occupational exposure refers to exposure of people who may not formally work on farms but live on or near sprayed areas or participate in unpaid farmwork [14].

Several studies have found that farmers in their majority did not respect good practices recommended by international agencies and local authorities. Exposure to pesticides therefore often occurs while preparing the spray solutions, loading in the spray tank, and applying the pesticide [15]. Circumstances of poisoning vary: 27% during spraying, 20% by ingestion (drinking, food contaminated by hands that have been used to manipulate pesticides), 13% occurred at home, 7% in the kitchen, and 3% during fishing [4].

Occupational exposure concerned up to 30% of poisoning cases. Common occupational risk factors include (i) absence of personal protective equipment, (ii) spraying in a direction opposite to wind, (iii) aerial sprayed, (iv) pesticides pouring on the skin, (v) pesticides entering in the eyes, noise, and mouth, and (vi) respect of the indicated dosage.

Putting together these risk factors can help develop an occupational and para-occupational exposure model, in order to estimate exposure circumstances for a rapid and adequate case management.

The gravity and the seriousness of exposures will also depend on the toxicological class of the pesticides, the quantity of pesticides exposed, the frequency, and other vulnerability factors.

In many country of the continent, diseases causes by pesticides are not recognized as professional diseases and include in security health assessment measures to be carried out at workplaces, especially where agriculture mainly depends on family farmers and small exploitations.

There are many different pesticide exposure scenarios. Operator exposure monitoring is of great concern, because it is known that operators receive more pesticide exposure than any other type of worker due to their close proximity and amount of pesticide handled.

Models give the possibilities to estimate the exposure to an active substance or to rank exposure of one pesticide to others used in similar conditions. Model development required a clear formulation of the problem, as well as proper selection of key variables and indicators. First pesticide exposures models were developed around 1980, and since then, these models had been constantly refined. It should be recalled that models proposed are only as accurate as the input values fed into them. These models in some extend are complementary as they deal with various aspect and routes of pesticides exposures. **Table 2** is a summary of some exposure models involving human monitoring.

Models	Exposure predicts	Reference
SHEDS	Post-application residential exposure	(https://www.epa.gov/pesticide-science-and-assessing-pesticide-risks/models-pesticide-risk-assessment)
EPA residential	Adult applicator, adult and child post-application indoors and outdoors oral, dermal, and inhalation	US EPA 2012 (https://www.epa.gov/pesticide-science-and-assessing-pesticide-risks/standard-operating-procedures-residential-pesticide)
EUROPOEM II	European predictive operator exposure model based on underlying operator exposure studies	Van Hemmen, 2001 (https://academic.oup.com/annweh/article-abstract/37/5/541/167727)
AHETF	Mixer/loader and applicator dermal and inhalation exposure under long pants and long sleeve shirt with protective gloves	(https://archive.epa.gov/pesticides/news/web/html/occ-exposure-data.html)
AOEM	Predictive model for estimating operator exposure proposed for use in the EU	Guidance on the assessment for exposure for operators, workers, residents, and bystanders in risk assessment for plant protection products (https://www.efsa.europa.eu/fr/efsajournal/pub/3874)
OPPED	Occupational post-application exposure (e.g., harvesting, weeding, etc.)	(https://www.epa.gov/pesticide-science-and-assessing-pesticide-risks/models-pesticide-risk-assessment)
Calendex ™	Aggregated exposure that incorporated the probability of simultaneous exposures across multiple pathways	US EPA
CARES©	Dietary and residential (surface, hand to mouth, air)	(http://caresng.org/)
CONSEXPO	Exposure to substances from consumers products that are used indoors	(https://www.rivm.nl/en/consexpo)
Lifeline	Cumulative and aggregate exposure to pesticides in foods, drinking water, and residential use	(https://www.thelifelinegroup.org/)

Table 2.
Summary of some pesticides exposure models involving human monitoring (adapted from [16]).

Assumptions and parameters used to build these models are often based on pesticide uses in European and North American models, which do not properly correspond to various exposure scenarios found in Africa. Agricultural pesticides in Africa is mostly used in family farming, and some related features of exposures remains to be taken into account to fit existing models to ongoing practices on the continent.

4. Farmers perceptions and experience of harmful effects of pesticides

Pesticides toxicity varies from one molecule to another. Classification by the World Health Organization (WHO) is done according to pesticides lethal dose 50 (LD50). LD50 varies from slightly toxic (LD50 > 5000 mg/kg) to extremely toxic (LD50 < 50 mg/kg). Depending on circumstances, exposure can be done through breathing, eating, or drinking or by contact with the skin or eyes. During an investigation by Cheke et al., a medical examination was done on farmers in Ekiti

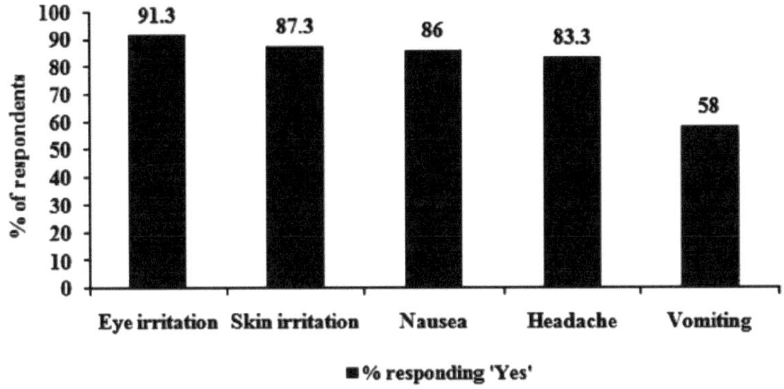

Figure 2.
Pesticide health symptoms experienced by farmers.

State, Nigeria, after taking their perceptions on any harmful effects they had experienced [17]. More than 91% of farmers reported to had suffered from pesticide-related health symptoms during or after applications. Symptoms were nausea, headache, vomiting, eye irritation, and skin problems, as shown in **Figure 2**.

In 2011, Mokhele determined the perceptions and awareness of farmworkers in Lesotho in South African Republic regarding the use of pesticides and the potential effects on their health. A total of 30 farmers from 6 farms participated [18]. The majority (85%) of farmworkers terminated their educational studies at the end of primary school. About 93% of farmworkers had received no training in the use of pesticides. A total of 52% of the farmworkers never wore rubber gloves when using or handling pesticides. All farmworkers in this study used a knapsack sprayer.

In the East African Rift (Ethiopia, Kenya, Tanzania, Uganda), Jacob de Boer et al. review literature about import, disposal, and health impacts of pesticides. They found out that in Ethiopia, few cases of poisoning have been reported [19]. Documented cases reported 81 pest control workers who were exposed to organochlorine pesticides (OCPs) (chlorpyrifos and profenofos) had lower cholinesterase levels after pesticide spraying. In Tanzania, data revealed that more than 50% of farmers have experienced headaches, excessive salivation, nausea, vomiting, and skin or eye irritation. Over 40% experienced dizziness, blurred vision, sleeplessness, and breathing difficulties, and over 20% reported tremors, diarrhea, chest pain, pain when urinating, fever, wheezing, or nosebleed.

In Kenya, Tsimbiri et al. testified that the main health impact of pesticides on residents and workers at Lake Naivasha in Kenya were headache and miserableness, followed by respiratory symptoms [20].

Apart from occupational and para-occupational exposures, pesticide dietary intake appears to be an overneglected issue. Even when good agricultural practices are followed, it is recognized that residues of pesticides remain on treated foods and agricultural products. The level of residues can be much higher in scenarios where pesticide misuses are numerous like in Africa.

5. Levels of pesticide residues in some highly consumed foods

In most African countries, pesticide residues are not of concerns in agricultural products sold at local markets, on the contrary to export products, which raises more attention. Residues on produce resulting from the inappropriate use of

N^0	Country	Pesticide/active ingredient	Class	Year of analysis	Commodity	Sampling location	Residu level (average) (mg/kg)	MRL (mg/kg)	MRL codex (mg/kg)	Reference
1	Ghana	Chlorpyrifos	Insecticide	2012	Fresh tomatoes	Kumasi six main markets	0.046 ± 0.01	0.5	1	[21]
2	Nigeria	p,p′-DDT (dichlorodiphényltrichloroéthane)	Anti-vectorial	2011	Tomatoes	Farms	0.046 ± 0.010	0. 1	—	[22]
3	Ghana	Cypermethrin	Insecticide	2012	Tomatoes	Kumasi six main markets	0.035 ± 0.005	0.5	0.2	[21]
4	Ghana	Diazinon	Insecticide	2012	Tomatoes	Kumasi six main markets	0.009 ± 0.003	0.5	0.5	[21]
5	Ghana	Dimethoate			Tomatoes	Kumasi six main markets	0.013 ± 0.009	0.02	NI	[21]
6	Ghana	Malathion	Insecticide		Tomatoes	Kumasi six main markets	0.038 ± 0.032	3.0	—	[21]
7	Egypt	Malathion	Insecticide	2014	Tomatoes	NI	0.025 ± 0.005	0.5	0.5	[23]
8	Egypt	L-Cyhalothrin	Insecticide	2014	Tomatoes	NI	0.070 ± 0.01	0. 1	—	[23]
9	Tanzania	Chlorpyrifos	Insecticide	2016	Tomatoes	—	7.528*	1.0	1	[24]
10	Tanzania	Ridomil	Insecticide	2016	Tomatoes	—	2854.729*	0.5	NI*	[24] et
11	South Africa	Chlorpyrifos	Insecticide	2016	Carrots	Johannesburg Supermarkets	0.02	0.05	0. 1	[25]
12	Nigeria	Lindane	Insecticide	2012	Beans	Lagos markets	0.192	—	NI	[26]
13	Nigeria	Malathion	Insecticide	2018	Rice	Ondo State	4.82 ± 0.001	—	NI	[27]
14	Nigeria	Carbofuran	Insecticide	2018	Rice	Ondo State	0.08 ± 0.001	—	1	[27]
15	Nigeria	Dichlorvos		2018	Wheat	—	0. 15 ±0.001	—	NI	[27]
16	Togo	Lindane	Insecticide	2012	Beans	—	0.01	—	—	[28]

N^0	Country	Pesticide/active ingredient	Class	Year of analysis	Commodity	Sampling location	Residu level (average) (mg/kg)	MRL (mg/kg)	MRL codex (mg/kg)	Reference
17	Ethiopia	Cypermethrin	Insecticide	2014	Red pepper	Local markets	0. 30*	—	0. 1	[29]
18	Ethiopia	Cypermethrin	Insecticide	2014	Maize	Local markets	0.156*	—	0.05	[29]
19	Cameroon	Carbofuran	Insecticide	2010	Maize	Markets of northern regions	0.001	—	0.05	[30]
20	Cameroon	Lindane	Insecticide	2010	Cowpea	Markets of northern regions	0. 1	—	—	[30]

NI: Not indicated

Table 3.
Pesticide residue levels in some of highly consumed food items.

pesticides are however one of the most important food safety concerns. **Table 3** gives a snapshot of pesticide residue levels in foods from the 10 countries identified as main importers.

Table 3 shows just a quick example of the type of pesticide residues that can be found in foods sold in market places of the continent. In all countries cited in the table, we have not been able to find any database of pesticide residues in foods. We focus on top 10 African countries importing pesticides from China. Others pesticide residues have been detected in African food products during their export into the European markets and notified on the Rapid Alert System for Food and Feed (RASFF) portal; an example of ethephon is found in pineapples coming from South Africa. In that specific case, the level of ethephon detected was **7.2 mg/kg and was treated as serious** [31].

Although herbicides are massively used, residues most often found in food items are insecticides. This can be due to their late uses in the production or during food conservation, especially in grains and cereals. Residues in most consumed foods are a clear indication of the level of exposure of the population.

Vegetables and grains appeared to be more frequently contaminated with residues.

In addition, multiple contaminations by 2–10 different pesticide active ingredients are also frequent, raising an issue of cocktail effects for a mixture of pesticides. Pesticides in combination may be far more dangerous and more serious threats for public health. Previous studies done on animal models showed various cocktail effects. Many chemical families of compound are involved specifically in synergy, addition, potentiation, and antagonism when combined.

Regulations of residues differ with countries; some countries had set their own maximum residue limits (MRLs) for certain pesticides in selected foods, and others have decided to adopt Codex alimentarius MRL. In both cases, actual levels of residues are found often to exceed national and/or international set MRLs.

6. Implications for pesticides dietary exposures and health risks

Pesticide residues refer to the pesticides that may remain on or in food after they are applied to food crops. Exposure of the general population to these residues most commonly occurs through consumption of treated commodities. Many food commodities are concerned with the exposure to several pesticide chemicals, especially staple foods, making it complex to estimate the dietary intake of residues.

To more accurately appreciate the health risk of pesticide residue intake, many approaches can be used. All of them are based on the assessment of level of residues in food commodities and the quantity of food consumed. Apart from being used as a starting point for estimating consumer exposures, MRLs are used also as reference points to decide misuses and as trade standards.

The use of food commodity MRL is a convenient way to assess the theoretical maximum daily intake (TMDI). This approach considered the set MRL as the contamination level assuming that farmers applied good agricultural practices. Unfortunately, from previous reading, we have seen that a huge number of farmers do not respect good agricultural practices, suggesting a higher level of residues and a higher risk for consumers.

In the example of **Table 3**, we observed that certain residue levels are higher than the set MRLs.

In South African Republic, Dalvie and London investigated the presence of pesticide residues in wheat produced and imported in the country and their health risks. Eight different pesticides were detected in total. The most frequently detected pesticides were mercaptothion (99%), permethrin (19%), and chlorpyrifos (17%).

Nine (11%) samples exceeded the EU wheat. Risk index calculated was found to be lower in more than half of cases [32].

Determining exposure values based on pesticide residue levels and the food consumption is also possible using either the deterministic or probabilistic approach. The deterministic approach is simpler and based on single-point estimates for each variable in the model. On the other hands, probabilistic approach allows using all possible values for each variable to be taken into account, and each possible model outcome is weighted by the probability of its occurrence. Probabilistic is therefore advantageous in that all available data are used, the exposure estimate is presented as a distribution, and variability and uncertainty can be quantified. The deterministic approach may be used as a screening tool to identify problematic pesticides, followed by the probabilistic approach to see if the point estimate actually gives rise to concerns.

In Tanzania, Kimanya et al., in 2016, estimated deterministically the dietary pesticide exposure of population to three pesticides through consumption of fresh tomato. Pesticide levels were detected for permethrin (mean, 5.2899 mg/kg), chlorpyrifos (mean, 7.5281 mg/kg), and ridomil (mean, 2854.279 mg/kg) in 18% of samples. Health risk indices, determined as ratio of estimated daily intake to acceptable daily exposure, for chlorpyrifos, permethrin, and ridomil were greater than one, which implies that lifetime consumption of fresh tomatoes can pose health risk for chlorpyrifos, permethrin, and ridomil for population of Meru District [33].

7. Conclusion and perspectives

Synthetic pesticides have been used in Africa since they were introduced by colonial masters. Over the years, the quantities of pesticides used especially in agriculture have exponentially increased. These products are not manufactured in African countries, but mainly imported from developing economies. Products were previously imported from Europe but, since few years, are growingly coming from China. Herbicides and insecticides constitute the top 20 of imported pesticides.

Although pesticides are homologated before they can be used, their supply to farmers followed many channels among which some that are illegal, particularly when it comes to supply products in rural settings.

Moreover, a lot of misuses are still observed on the field. Good agricultural practices are not known by a great number of farmers and workers, increasing therefore risk factors of exposures and poisoning.

Farmers are frequently subjected to multiple exposures at workplace (inhalation, skin contact), while populations are exposed to pesticide residues found in the environment and accumulated in foods commodities. Estimating the level of human exposure for occupational and non-occupational scenarios remains a real challenge for risk assessors. A number of tools and techniques exist but do not fit to all situations, particularly in cases of misuses of pesticide applications. Predictive models have been designed to quickly capture and assess exposure cases, but these models remain to be improved to fit unusual practices by small farmers in rural areas of Africa.

Farmers are all aware of risks that pesticides can cause to their health; however, they remain reluctant in using protective equipment and to adopt preventive behavior. Most of them have reported to experience at workplaces symptoms of pesticide poisoning (eye irritation, skin irritation, nausea, headache, and vomiting).

Because of the absence of toxicovigilance systems, as well as surveillance and monitoring program, very few cases of pesticides poisoning have been documented. Epidemiological data are not known. Actual risk assessment is based on the estimation of dietary intake of pesticides. These calculations showed frequent contamination of food items by pesticide residues, especially insecticides. The level of residues is often higher than national and international maximum residue limits, suggesting real public health threats for the whole population. In addition, cocktail and cumulative effects of multiple residues remain to be investigated.

Author details

Pouokam Guy Bertrand[1,2]*

1 Nutrition and Food Safety and Wholesomeness, Cameroon

2 Laboratory of Biochemistry, Faculty of Medicine and Biomedical Sciences, University of Yaoundé 1, Cameroon

*Address all correspondence to: guy.pouokam@noodlesonlus.org

References

[1] Wandiga SO. Use and distribution of organochlorine pesticides. The future in Africa. Pure and Applied Chemistry. 2001;**73**(7):1147-1155

[2] FAOSTAT: Pesticides Trade Database. Available from: http://www.fao.org/faostat/en/#data/RT [Accesed: December 17, 2018]

[3] Erwin X. Briefing on Export of Pesticides to Africa From China in 2015. Available from: http://news.agropages.com/News/NewsDetail---17241.htm [Accessed: November 11, 2018]

[4] Pouokam GB, Lemnyuy Album W, Ndikontar AS, MEH S. Pilot study in Cameroon to understand safe uses of pesticides in agriculture, risk factors for farmers' exposure and management of accidental cases. Toxics. 2017;**5**(4):E30. DOI: 10.3390/toxics5040030

[5] Quinn LP, de Vos BJ, Fernandes-Whaley M, Roos C, Bouwman H, Kylin H, et al. Pesticide use in South Africa: One of the largest importers of pesticides in Africa. In: Stoytcheva M, editor. Pesticides in the Modern World - Pesticides Use and Management. InTech; 2011. ISBN: 978-953-307- 459-7. Available from: http://www.intechopen.com/books/pesticides-in-the-modern-world-pesticidesuse-and-management/pesticide-use-in-south-africa-one-of-the-largest-importers-of-pesticides-in-africa

[6] Van den Brink PJ, Kwakye MO, Mengistie B, Ofosu-Anim J, Nuer ATK. Pesticide registration, distribution and use practices in Ghana. Environment, Development and Sustainability. 2018. DOI: 10.1007/s10668-018-0154-7

[7] CILSS: Liste Globale des Pesticides Autorisés par le Comité Sahélien des Pesticides dans les pays Membres du CILSS en Novembre 2017. Available from: https://www.csan-niger.com/pesticides-homologues-novembre-2017.php [Accessed: November 20, 2018]

[8] Megeed MA. Pesticides Management in Egypt. Agricultural Pesticide Committee (APC), Ministry of Agriculture and Land Reclamation; Published online 24 Mar 2017. DOI: 10.3390/ijerph14040340

[9] Tanzanian National Audit Office. A Report on the Performance Audit Report on the Management of Pesticides in Agriculture. The United Republic of Tanzania National Audit Office. Available from: http://www.nao.go.tz/?wpfb_dl=264; 2018

[10] USAID task order AID #663-C-17-00001: Pesticide Evaluation Report and Safer Use Action Plan (PERSUAP). Available from: https://ecd.usaid.gov/repository/doc/50680.docx; 2017. p. 38

[11] AU–IAPSC/Flaubert Nana Sani. Pesticides utilization in Africa: Status and trends. In: Presentation During a Workshop on Harmonization of Pesticides Regulations in Africa; 17-19 April 2016; Cairo, Egypt; 2000

[12] Bon H, Huat J, Parrot L, Sinzogan A, Martin T, Malezieux E, et al. Pesticides risks from fruits and vegetable pest management by small farmers in sub-saharan Africa. A review. Agronomy for Sustainable Development. 2015, 2014;**34**(4):723-736

[13] Kumar N, Pathera AK, Saini P, Kumar M. Harmful effects of pesticides on human health. Annals of Agri Bio Research. 2012;**17**(2):125-127

[14] Blewett C, Nicol A-M. Para-occupational exposure to pesticides in agricultural families: Developing models for risk assessment. Occupational Environmental Medicine. 2011;**68**(Suppl):A107-A

[15] Dahab AA, Jallow MFA, Albaho MS. Evaluation of pesticides residues in some Egyptian fruits. In: Environmental and Human Health Impacts of Pesticide Use in Agriculture. International Journal of Environmental Research and Public Health; Aug 2017;**14**(8):833

[16] Ross J, Driver J, Lunchick C, O'Mahony C. Models for estimating human exposure to pesticides. Outlooks on Pest Management. 2015;**26**(1):33-37. DOI: 10.1564/v26_feb_09

[17] Cheke RA, Oluwole O. Health and environmental impacts of pesticide use practices: A case study of farmers in Ekiti State, Nigeria. International Journal in Agricultural Sustainability. 2009;**3**:153-163

[18] Mokhele TA. Potential health effects of pesticide use on farmworkers in Lesotho. South African Journal of Science. 2011;**107**(7-8)

[19] de Boer J, MuluLoha K, Lamoree M, Weiss JM. Import, disposal, and health impacts of pesticides in the East Africa Rift(EAR) zone: A review on management and policy analysis. Crop Protection. 2018;**112**:322-331

[20] Tsimbiri PF, Moturi WN, Sawe J, Henley P, Bend JR. Health impact of pesticides on residents and horticultural workers in the Lake Naivasha Region. Occupational Diseases and Environmental Medicine. 2015;**3**:24-34. DOI: 10.4236/odem.2015.32004

[21] Crentsil BK, Buah-Kwofie A, Enimil E, Agyei-Martey G. Residues of organochlorine pesticides in vegetables marketed in greater Accra region–Ghana. Food Control. 2012;**25**(2):537-542

[22] Benson N, Olufunke A. Assessment of contamination by organochlorine pesticides in Solanum lycopersicum L. and Capsicum annuum L.: A market survey in Nigeria. African Journal of Environmental Science and Technology. 2011;**5**(6):437-442

[23] Gad Alla SA, Almaz MM, Thabet WM, Nabil MM. Evaluation of pesticide residues in some Egyptian fruits. International Journal of Environment. 2015;**04**(01):87-97

[24] Victoria K, Kassim N, Kimanya M. Risk of exposures of pesticide residues from tomato in Tanzania. African Journal of Food Sciences. 2017;**11**(8):255-262

[25] Korsten L, Thomas M, Chidamba L. Monitoring pesticide residues in fruits and vegetables at two of the biggest fresh produce markets in Africa. Journal of Food Protection. 2016;**79**(11):1938-1945

[26] Gwary OM, Hati SS, Dimariand GA, Ogugbuaja VO. Pesticide residues in bean samples from Northeastern Nigeria. ARPN Journal of Science and Technology. 2012;**2**(2):79-84

[27] Akinneye Joseph O, Adedolapo AO, Adesina Femi P. Quantification of organophosphate and carbamate residue on stored grains in Ondo State, Nigeria. Journal of Biology and Medicine. 2018;**2**(1):001-006. DOI: 10.17352/aur.000003

[28] Akpagana K, Kanda M, Djaneye-Boundjou G, Wala K, Gnandi K, Batawila K, et al. Assessment of pesticide residues and trace element contamination in market gardens of Togo. African Journal of Environmental Science and Technology. October 2012;**6**(10):380-390

[29] Mekonen S, Ambelu A, Spanoghe P. Pesticide residue evaluation in major staple food items of Ethiopia using the Quechers method: A case study from the Jimma zone. Environmental Toxicology and Chemistry. 2014;**33**(6):1294-1302

[30] Sonchieu J, Benoit Ngassoum M, Bosco Tchatchueng J, Srivastava AK,

Srivastava LP. Survey of pesticide residues in maize, cowpea and millet from northern Cameroon: Part I. Food Additives and Contaminants: Part B. 2010;**3**(3):178-184

[31] RASFF Portal. Available from: https://webgate.ec.europa.eu/rasff-window/portal/?event=notificationDetail&NOTIF_REFERENCE=2014.0800

[32] Dalvie MA, London L. Risk assessment of pesticide residues in South African raw wheat. Crop Protection. 2009;**28**(10):864-869

[33] Kimanya M, Kariathi V, Kassim N. Pesticide exposure from fresh tomatoes and its relationship with pesticide application practices in Meru district. Cogent Food & Agriculture. 2016;**2**:1196808

Chapter 6

Pesticides, Anthropogenic Activities, History and the Health of Our Environment: Lessons from Africa

Wilbert Bunini Manyilizu

Abstract

This chapter describes the historical events related to pesticide use. The description of events focuses on human activities that necessitated the use of chemical agents for pest control to protect crops, and animals including humans in African countries. The description covers the common pests in Africa and the need for pest control using pesticides. History of pesticide use in Africa and the ban of organochlorines are covered. Controversies under discussion in Africa and the current trend of pesticide use in Africa are part of the chapter as well as pesticide import and supply. Hazard and risk of exposure of biological organisms including humans to pesticides due to anthropogenic activities in Africa and pros and cons of pesticide use in Africa are covered.

Keywords: anthropogenic activities, history of pesticides, humans, environment, health, Africa

1. Introduction

This chapter describes the historical events related to pesticide use and the pros and cons of pesticide use in Africa. Description for *pesticide* as any chemical used to prevent, destroy, or repel pests and also the description of *pest* as any species that interferes with human activities, properties, or health have been provided with examples. In Africa, rapid population growth, illiteracy, food insecurity, weak control systems, and poverty have accelerated the use and misuse of pesticides. Based on the latest, 2018, United Nations estimates the current population of Africa is now estimated at 1.3 billion, that is, 16.6% of the total global population. A large part of African population lives in tropical and subtropical climate with high temperatures [1] and moisture favorable for insects' population growth, as well as movement, agricultural and animal husbandry activities throughout the year [2]. Through these dynamics, humans modify the components of disease agents, including moisture to promote disease occurrence and spread. Hunger and malnutrition, as a result, are affecting many regions in Africa. In 2016, FAO estimated that 27.4% of the population in Africa is affected by severe food insecurity. Since food insecurity is on the rise, especially in sub-Saharan Africa, the need for increase of food productivity and use of pesticides are unavoidable. Over the past decades, the history of pesticides for agriculture, public health, and construction industry in Africa has gone through milestones with several challenges. These challenges range from limited control in import, distribution, use, storage, and disposal of pesticides. As a result, the risk of exposure and health impact to humans and environment has become another challenge. This chapter not only describes the trend of pesticide use and the negative consequences experienced in the past and the current status but also predicts the future implications for environment and health. Controversies regarding the benefits of pesticide use and the disadvantages that are magnified by lack of knowledge, protection, and malpractice with pesticides are highlighted.

2. Anthropogenic activities

Human activities are a part of struggle for meeting basic needs of life. In order for humans to sustain life, they must discover better means for addressing the development challenges including those relating food security and safety for a peaceful and secure life. In order to sustain productivity, food security, and safety for survival and growth, humans have to control the environmental challenges due to anthropogenic activity including nuisance and threats.

Since before 2000 BC, humans have been utilizing pesticides to protect crops. In Mesopotamia, about 4500 years ago, they used elemental sulfur dusting as pesticide for their crops. In other places, they used poisonous plants for pest control [3]. Other methods of pest control included burning grasses not only to kill insects and to control plant diseases but also to inhibit the growth of unwanted weeds. The serious use of pesticides in the agriculture started in the nineteenth century and expanded in the twentieth century [4]. Pesticides were used to control various pests and disease carriers, such as mosquitoes, fleas, ticks, mice, and rats.

Use of pesticides to control pests of importance in public health and agriculture including animal husbandry and poultry has been necessary for improving health as well as quantity and quality of yield for feeding the growing population. As a result, these pesticides reach the untargeted organisms through direct contact, polluted water sources, air, soil, and the food chain due to weak control systems for importation, supply, use, and disposal. In general, human activities that involve application of pesticides pollute and destroy habitats, untargeted animals, and some plant species. Thus, as unwanted effect, exposure of pests to pesticides leads to pest resistance problem, loss of many untargeted species, and also biological magnification through food chain.

3. Pests in Africa

Despite the fact that poorly controlled human activities threaten different untargeted species, agriculture in Africa is threatened by pests, including insects. The insects can either be endemic or epidemic. The endemic insects in Africa include cereal stalk borers that destroy different kinds of cereal crops and crop-eating fall armyworms that destroy a wide variety of crops and also whiteflies that destroy root/tuber crops (e.g., cassava is one of main sources of carbohydrates). Bean flies, aphids, thrips, leafhoppers, whiteflies, and leaf beetles are also among common and endemic insects that destroy legume crops' source of protein and many more insects in Africa [5].

Epidemic insect attacks in Africa include locust outbreaks (e.g., Madagascar in 1997) that inflicted severe damage to crops and cattle pastures around the country. In this locust outbreak control, fipronil (insecticide) was donated by developed countries, later impact evaluation reported detrimental fipronil effects, ranging from genotoxicity and cytotoxicity, and impaired immune function, to reduced growth and reproductive success of vertebrates, often at concentrations below that which is associated with mortality [6].

Other pests include fungi, virus, and bacteria. There are substantial estimated losses caused by these pests per year. For example, in Tanzania, economic damage due to the other pests on crop productivity is estimated at 50% (Controller and Auditor General established that in 2015).

Human life in Africa is also threatened by vector-borne diseases. Such vectors (pests) transmitting diseases include female anopheles mosquito that transmits *Plasmodium falciparum* causing malaria. Culex and other mosquito species transmit *Wuchereria bancrofti* (mostly) causing elephantiasis leading to permanent disability. Fleas harbored by rats transmit *Yersinia pestis* causing plague and tsetse flies transmit *Trypanosoma brucei rhodesiense* and *Trypanosoma brucei gambiense* to cause sleeping sickness. Fresh water snails transmit schistosomes causing intestinal and urinary schistosomiasis [7–9]. All these cause a lot of socioeconomic losses due to diseases and deaths they cause to humans. Other pests like ticks cause health problems to animals.

4. Need for pest control

In order to achieve human and animal health and other social and economic targets, humans need to control pests, so that they can reduce nuisance to increase quantity and quality of crop harvest, the value of harvested crops for sale, and livestock. Not only pest control is necessary for decreasing human and animal diseases and deaths but also for decreasing nuisance, direct destructions of properties as well as promoting peace for social and economic activities to occur.

5. Pesticides used in Africa

Pesticides that are mostly used in Africa include insecticides (insects), fungicides (fungi), acaricide (ticks, mites), antibiotics (bacteria), molluscicide (snails),

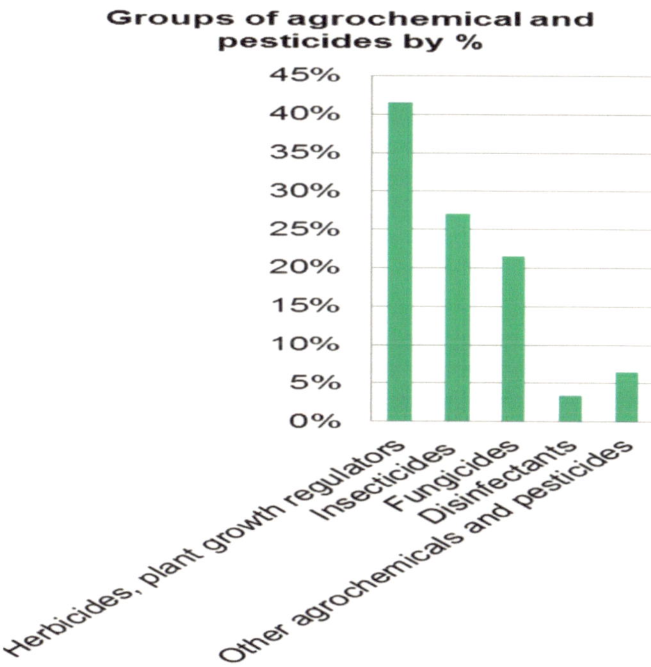

Figure 1.
Pesticides and other agrochemicals used in Africa.

nematicide (nematodes), ovicide (birds), repellents (vectors), rodenticides (rodents), and herbicides (weeds) (**Figure 1**).

6. General history of pesticides

Initial history of pesticides is well documented in countries outside Africa. For example, the first generation of pesticides contained naturally occurring metal elements. These inorganic metals included lead, calcium, arsenic, and mercury. These pesticides were mainly discovered and used in European countries and the USA. Later, they were found to be less effective for insect control and they were highly toxic to plants and animals [10]. No data are available on the use of first generation of pesticides in Africa.

Data available for pesticide use in Africa are on second generation of pesticides (organochlorines). Second-generation (synthetic) pesticides were organochlorines such as dichlorodiphenyltrichloroethane (DDT). DDT was first synthesized in 1874 by the young Austrian chemist Othmar Zeidler (as a doctoral student), but in 1939, the DDT's insecticidal action was first discovered by the Swiss chemist Paul Hermann Müller [11]. These organochlorines replaced inorganic pesticides (first generation). Then, a third generation of pesticides included organophosphates, carbamates, pyrethroids, etc. These are the currently used pesticides in Africa and elsewhere, they were introduced between 1960s and 1980s [12, 13].

7. History of pesticide use in Africa

In Africa, data on pesticide use [14] are available from when (1939–1960s) the second-generation organochlorines were reported. In Tanzania, DDT was introduced during WWII for malaria and typhus and later, after WWII, it was available for public health and farm vector control. In 1945, DDT was introduced in Monrovia, Liberia, for indoor residual spraying (IRS) for controlling malaria vector [15]. After WWII, there was effective worldwide marketing and from 1950s and after that, there came introduction of lindane, dieldrin, chlordane, and endosulfan. DDT brought happiness to many countries because it was a broad-spectrum pesticide effective at killing pests and could be used by inexperienced people, improved crop yields, and needed no re-application—so, it was a cost-saving pesticide.

8. Advantages of DDT

During WW II, DDT was applied to control lice (typhus) that caused typhus fever [16], and to control mosquito that caused malaria [17]. Significant decline of malaria transmission and deaths after use of DDT was reported in different African countries from 1940s to 1950s on.

9. Negative effects of DDT

Later, research showed that DDT had a negative effect on the environment and biodiversity. Rachel Carson published the book, namely, Silent Spring in 1962 and the message from this book was an eye opener. She reported that DDT thinned bird egg shells, and, that, eggs were not able to support the weight of incubating birds, so not able to hatch. The reason was that the egg shells lacked enough calcium due to DDT. In addition, DDT had estrogenic effect, thus affecting reproduction. DDT was affecting the nervous system and it also affected immunity leading to failure to resist against infections in animals. So, it was a threat to extinction of birds and other wild creatures [18]. A number of studies have revealed DDT residues in many kinds of samples in several African countries like Nigeria, Tunisia, Ethiopia, Burundi, South Africa, etc., in plants, animal feed, livestock and wild animals, birds including chicken, fish, and humans [19].

In the food chain, plants might have low DDT residues, and they are eaten by chicken, fish, and animals; the DDT concentration levels increase in the tissues, and high up in the chain to reach even innocent newborns via contaminated breast milk (i.e., highest DDT concentration level in the food chain).

As a result, examples of literature about negative effects of DDT in humans in Africa include those in breast milk. Organochlorine pesticides (OCPs) were reported to be present in human breast milk, thus causing health risk to nursing infants in northern Tanzania in 2017 by Müller et al. [20]; in South Africa in 2006 by Bouwman et al. [21]; and also reported in milk and serum of Ghanaian farmers [22]. Furthermore, long-term effects of DDT exposure not only affected semen, fertility, and sexual function of farm workers in South Africa [23], but also caused DDT genotoxicity to cultured lymphocytes in Tunisia [24] and reduced half-life of paracetamol in highly exposed mothers in Zimbabwe [25]. Paracetamol is useful for fever and different kinds of pains in humans, its half-life is 1–3 h (prescription is after minimum of 6 h). Say the half-life is reduced to 30 min due to DDT residues in the body, then, the interval of taking paracetamol must be less than 6 h. Thus, paracetamol toxicity to liver is increased due to increased frequency or it becomes a useless drug in places where DDT is applied.

10. Ban of organochlorines (OCs or OCPs)

The negative effects of OCs to the environment and humans outweighed their benefits, leading to ban of OC pesticides; reasons included resistance to degradation in nature and living organisms, its toxicity to biodiversity including humans but also pests developed resistance. So, it was banned in developed countries in 1970s (Europe and the USA).

10.1 Stockholm convention

A worldwide ban on production and use was formalized under the Stockholm Convention on Persistent Organic Pollutants (POPs) signed in 2001 and effected in May 2004. The ban included DDT and other 11 persistent organic pollutants (POPs), namely the dirty dozen. The dirty dozen are characterized by persistence due to slow degradation, they are lipophilic (i.e., high affinity for fatty tissue); so, they accumulate in fatty tissues of living organisms (bioaccumulation), and then there is an increase of concentration in food chain (biomagnification). Due to long (persistence) half-life, they can be transported far from the point of application via air [26]. DDT may be transported from tropical countries to polar regions via evaporation, then, condensed and in summer, again, they can evaporate (grasshopper effect).

11. Controversy under discussion

DDT is the most effective pesticide for malaria control. Following the ban of the dirty dozen, should DDT be banned for every activity? In 1990, African countries tried

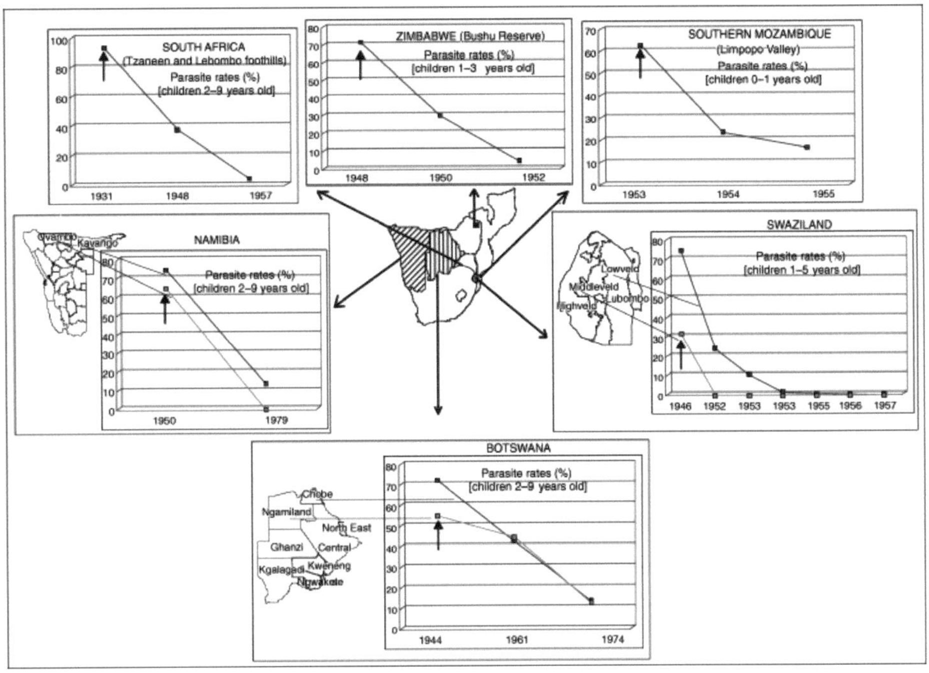

Figure 2.
Decline of malaria parasite rates in Africa after use of DDT.

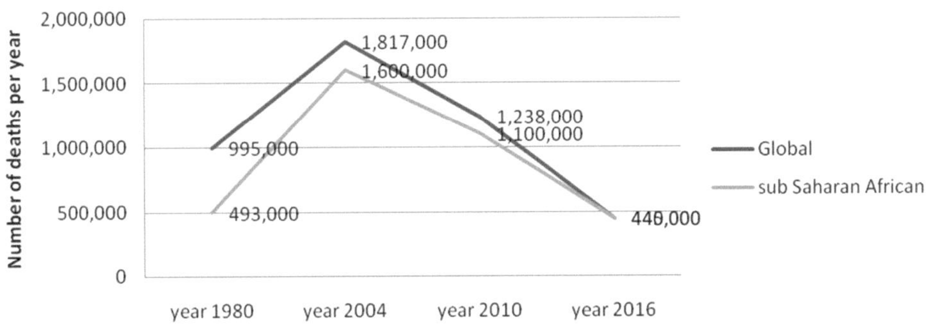

Figure 3.
Mortality due to malaria in Africa from 1980 to 2016 (90% contribution to global mortality). Figure based on data from UNICEF and WHO.

to substitute DDT with pyrethroids. The result was the rebound of malaria morbidity and mortality. So, the WHO allowed reintroduction of DDT in 2004. After 2004, mortality decreased. DDT use is reported to have led to decline of malaria morbidity and mortality in Africa. The challenge now for Africa is to rely on DDT use for malaria control despite the negative consequences, including the potential pest resistance.

Figure 2 has been adapted from European Journal TMIH by Musawenkosi et al., 2004. Historical review of malarial control in southern Africa with emphasis on the use of indoor residual house-spraying is given in **Figure 3**.

12. Trend of pesticide use in Africa

The organization currently known as African Union launched a program for self-sufficiency in agricultural food production in 1983. It was a 10-year program up to

1993. Example of Tanzania as one of the program implementers in Africa removed restrictions on imports of different pesticides including the banned pesticides; from there, was an observed rapid increase of pesticide imports. In addition to that, there was another increase of pesticide imports from 2000 to 2003 (about 5 times) [27].

13. Pesticide import and supply

On top of increased imports for self-reliance on food production through direct purchase, Africa also received donations from Europe (EU, UK), the USA, Asia (China and India), etc. [28]. Within the country, there are suppliers and distributors, that is registered companies and small-scale traders operating via local shops and also vendors.

14. Historical burden in African countries

As a result of donations and poorly planned imports, many countries in Africa have remained with obsolete pesticides accumulated over the past decades. These persistent organochlorine pesticides were stocked for use, but no longer useful, they required disposal because they have become a source of pollution to the environment and food chain and direct threat to human health.

15. Why did these stockpiles of pesticides end up in Africa?

Organochlorine pesticides were banned in developed countries in 1970s. At that time, many African countries received large donations of DDT and malathion for malaria control programs. Additional description was that the donations were for preparedness against locust outbreaks. This was an act of disposal and smart donation to solve environmental problems in donor countries. In addition, in 1991, during implementation of self-reliance program on food production, 1900 tons of banned pesticides manufactured in the USA were shipped to Africa. In some cases, there were even excessive donations without examining the actual need for these products in the recipient country. There were no prior arrangements for distribution and storage of these pesticides.

This was reported in Pesticides and agrochemical industry in sub-Saharan Africa, July 1994 (Contractual work prepared for division of food, agriculture and resource analysis-office of analysis, research and technical support bureau for Africa).

16. Is this a hazard to human health and environment?

Despite the cleanup program, there are still high levels of DDT and HCHs that were found in soil and water around. Although the visible remains of pesticides were removed, the soil is itself hazardous waste. For example, in Vikuge, in Tanzania, concentrations of DDT in grasses from nearby Vikuge were far above the acceptable limits for animal feed. Even at 6-km distance from Vikuge, DDT concentrations in grasses (animal feed) were still two times higher than the acceptable limits [29].

17. Africa stockpiles program funded by World Bank and non-bank sources

Other areas under the same cleanup program in 2013–2015 included Mali, Ethiopia, Morocco, Tunisia, and South Africa. Example of operations in Tanzania identified 14 sites of obsolete stockpiles. Three hundred tons of DDT of contaminated soils in Morogoro Region (one of the regions in Tanzania) were collected for destruction and 200 tons of DDT collected from the government-owned livestock farm at Vikuge [30].

Recent findings of pesticides (2016) in chicken eggs from Arusha, in Tanzania, by Polder et al. report that there are POPs including pesticides from free-ranging chicken eggs (free-ranging chicken are common in Africa for family use and for sale). They collect food from soil around the homes and come back during sunset. These findings from Arusha revealed extremely high levels of dieldrin in eggs from one specific urban farmer. This finding may reflect a possible source from an obsolete stockpile that was situated on that site before the town expanded.

18. Current pesticide use in Africa

Third generation of pesticides came in between 1960s and 1980s: these included organophosphates, carbamates, etc. These are the rapidly degraded pesticides, so they are less persistent in environment. They are acutely toxic to pests and more lethal in low dose compared to the banned organochlorines. The current global consumption of pesticides is at 2 million tons per year; of these, 25% (500,000 tons) is consumed in developing countries and 4% (80,000 tons) of global consumption is from Africa [31].

19. What is the problem of Africa with use of pesticides?

A survey report named Pesticides and Poverty [28] showed a number of problems noted in Africa; these include weak government organs for pesticide control systems, in particular, planning imports (imports may be in excess of requirements, so there is lack of efficiency). There are also weaknesses in supply and distribution (farmers accessed late and sometimes not according to needs). There is poor control (illegal entry of 2% pesticides, loop holes for misuse) and poor disposal plans of the remains. There is illegal trading (unwanted pesticides including WHO class I and unknown ingredients).

There are problems by users. These include not only improper practices (no personal protective equipment, contaminated water sources during pesticide applications) but also improper storage and disposal (throwing and burying containers in fields). In addition, users have low knowledge (on safe use and the associated health risks, also users cannot diagnose the plant disease and prescribe accurately), skills, and capacity. Some pesticide users have never attended formal education in school, these are the majority of that improperly use, store and dispose the pesticides.

20. Human exposure to pesticides in Africa

Common human exposure is through spraying (including mixing and loading), weeding, pruning, harvesting, etc., but also drift near the areas of pesticide applications, indoor spaying of mosquitoes, cockroaches, flies, ants, etc. Direct contact with contaminated materials (at farm/home) and ingestion (poor hygiene) are common exposure pathways.

21. What has been done with the problems?

African countries are taking some steps to address the described problems, these steps include the following: African member states have ratified the UN pesticide conventions and protocols as described by Flaubert Nana Sani (AU-IAPSC). Most African member states at the moment have Pesticide Evaluation Report & Safer Use Action Plans. At the moment, there are subregional regulatory bodies in Africa, these include: Central Africa Inter-State Pesticides Committee, the South East Africa Regulatory Committee on Harmonization, and another one for the west African member states. Among the activities that have been done by these member states at different levels include establishing harmonized pesticide registration, procedures, and evaluation criteria.

22. Pros and cons of pesticide use in Africa

22.1 Pros of pesticide use

Pesticides are important for economic development, food security (enough food, to avoid hunger), food production (able to conduct agricultural activities for food availability without pest disturbance), food safety (preventing biological harm to consumers), food quality (nutrients, appearance, texture, flavor, chemical, physical, microbial properties, etc.), vector disease control, improving human and animal health, decreasing morbidities and mortalities, insect nuisance control, and increased life quality. All these lead to peace.

Not only have controlled vector-related diseases (acute and chronic) including malaria morbidity and mortality been significantly reduced but also threat of elephantiasis and bilharziasis has gone down among many pest and vector-borne/related diseases. The same controlled picture is observed in animal health and zoonotic diseases.

22.2 Cons of pesticide use

Pesticides contaminate water, air, and soil, leading to damage of ecosystems (some organisms and habitats are destroyed and no longer exist in their natural habitat). Thus, pesticides diminish biodiversity (some biological species become extinct) and affect natural biological equilibrium. In affected systems of living organisms, some biological species are forced to live in new environment; thus, they adapt and may become pests. Some pests prevail in excess or less where not expected.

Other problems due to pesticides include pesticide resistance and costs to controlling resistance. Human and animal exposure to pesticides end into health problems and also it is reported that efficacy of the vaccine is reduced due to exposure to pesticides. Pesticides are threats to human health by directly causing diseases. Diseases due to pesticides can be divided into two kinds of manifestations: acute and chronic poisoning.

In acute poisoning (high dose), the body reacts to present with diarrhea, vomiting, coughing, difficult breathing, skin irritation, rashes, fasciculation, headache, dizziness, etc. When humans are exposed to low dose, chronic symptoms manifest including slow onset of symptoms. Pesticides are neurotoxicants, so they affect the central nervous system and manifest through loss of memory, orientation to time and space, etc., and on the peripheral nervous system, numbness of feet and hands manifest among other symptoms. Pesticides also have effects on reproductive system because they are endocrine disruptors (they affect reproduction, e.g., lead to abortions, etc.).

On immune system, they disturb body function, so resistance to infections is reduced. Cancers, for example, lymphomas, sarcoma, etc., are also reported to occur more in populations exposed to pesticides (cause effect not established). In addition, more than 40% of the health care professionals interviewed could not recognize pesticide poisoning cases; this reflects that the recognition for chronic manifestation of low-dose occupational exposure to pesticides may be worse [14].

Contamination of water threatens aquatic organisms, frogs, and fish, leading to extinction of aquatic biodiversity. Contamination of soil may lead to extinction of fertilizing organisms, whereas air pollution leads to population decline of pollinators (honey bees). The persistent organic pollutants are transported far away from the area of application and they end up in biomagnification that threatens human health.

In addition, environmental contamination causes lack of safe water supply for human consumption, which threatens human health. Contaminated water and grass lead to wildlife poisoning and extinction of wildlife including birds, leading to loss of small mammals, bird species, and insects.

Pesticides are expensive. Since pesticide use is a solution for temporary protection, it forces frequent use that in turn increases risk of exposures. As a result, pesticide resistance occurs. Pest resistance is a big challenge. When the pests do not die following pesticide application, the users increase the dose. At the same time, the natural enemies for pests are killed by pesticides during applications, leading to pest resurgence. Pests come back stronger than before because there is no natural enemy. To control pest resurgence, the users increase quantity and frequency for spraying. As a result, secondary pest outbreak occurs, in which, normal species become pests because natural enemy is destroyed. This kind of new pest is sprayed like target pest.

23. Conclusions and recommendations

Human struggle for survival has led to increased use of pesticides. Loopholes in controlling use and disposal of pesticides have threatened the human and environment health over decades. As a result, morbidities and mortalities and other negative consequences to untargeted biological organisms need serious considerations and adequate actions. Recommendations may not be limited to adjustment of the laws and regulations to be in harmony with international conventions and standards but also strengthening implementation and enforcement of the existing rules and regulations, registration and quality controls. Having infrastructure for handling sewage systems and proper disposal systems for pesticides and other chemicals and development of alternative for sustainable food production is important. Education and training on safe pesticide use, storage, disposition of pesticides and training in schools on environmental, occupational and dietary-related non-communicable diseases are necessary.

Acknowledgements

The following colleagues from Norwegian University of Life Sciences, Oslo, are sincerely acknowledged for their professional contributions that went into this piece of work. Prof Jan Ludvig Lyche (PhD) is acknowledged for critical reading and comments on the content. Dr. Anuschka Polder (PhD) is acknowledged for contribution on the flow of arguments during writing process. Dr. Mette Hellen Bjorge Müller (PhD) is acknowledged for constructive advices and comments.

Conflict of interest

It is declared that there is no conflict of interest between the author and any other part regarding the content of this chapter.

Author details

Wilbert Bunini Manyilizu
Department of Health Systems Management, Mzumbe University, Morogoro, Tanzania, East Africa

*Address all correspondence to: wmanyilizu@gmail.com

References

[1] Lunde TM, Lindtjorn B. Cattle and climate in Africa. How climate variability has influenced national cattle holdings from 1961-2008. PeerJ. 2013;**1**:e55

[2] Biber-Freudenberger L, Ziemacki J, Tonnang HE, Borgemeister C. Future risks of pest species under changing climatic conditions. PLoS One. 2016;**11**(4):e0153237

[3] http://pm22100.net/01_PDF_THEMES/wiki/PESTICIDES.pdf [Accessed: November 2, 2018]

[4] Jarman WM, Ballschmiter K. From coal to DDT. The history of the development of the pesticide DDT from synthetic dyes till silent spring. Endeavour. 2012;**36**(4):131-142

[5] Pretty J, Bharucha ZP. Integrated pest management for sustainable intensification of agriculture in Asia and Africa. Insects. 2015;**6**(1):152-182

[6] Gibbons D, Morrissey C, Mineau P. A review of the direct and indirect effects of neonicotinoids and fipronil on vertebrate wildlife. Environmental Science and Pollution Research International. 2015;**22**(1):103-118

[7] Ratmanov P, Mediannikov O, Raoult D. Vectorborne diseases in West Africa: Geographic distribution and geospatial characteristics. Transactions of the Royal Society of Tropical Medicine and Hygiene. 2013;**107**(5):273-284

[8] Githeko AK, Lindsay SW, Confalonieri UE, Patz JA. Climate change and vector-borne diseases: A regional analysis. Bulletin of the World Health Organization. 2000;**78**(9):1136-1147

[9] Asmare K, Abayneh T, Sibhat B, Shiferaw D, Szonyi B, Krontveit RI, et al. Major vectors and vector-borne diseases in small ruminants in Ethiopia: A systematic review. Acta Tropica. 2017;**170**:95-104

[10] http://soils.tfrec.wsu.edu/leadhistory.htm [Accessed: November 2, 2018]

[11] Harada T, Takeda M, Kojima S, Tomiyama N. Toxicity and carcinogenicity of dichlorodiphenyltrichloroethane (DDT). Toxicological Research. 2016;**32**(1):21-33

[12] Williams CM. Third-generation pesticides. Scientific American. 1967;**217**(1):13-17

[13] Vinson SB, Plapp FW Jr. Third generation pesticides: The potential for the development of resistance by insects. Journal of Agricultural and Food Chemistry. 1974;**22**(3):356-360

[14] Mbakaya CF, Ohayo-Mitoko GJ, Ngowi VA, Mbabazi R, Simwa JM, Maeda DN, et al. The status of pesticide usage in East Africa. African Journal of Health Sciences. 1994;**1**(1):37-41

[15] Webb JL Jr. The first large-scale use of synthetic insecticide for malaria control in tropical Africa: Lessons from Liberia, 1945-1962. Journal of the History of Medicine and Allied Sciences. 2011;**66**(3):347-376

[16] Snyder JC. Typhus fever in the second world war. California Medicine. 1947;**66**(1):3-10

[17] Bishopp FC. Insect problems in world war II with special references to the insecticide DDT. American Journal of Public Health and the Nation's Health. 1945;**35**(4):373-378

[18] Kabasenche WP, Skinner MK. DDT, epigenetic harm, and transgenerational

environmental justice. Environmental Health. 2014;**13**:62

[19] Snow RW, Amratia P, Kabaria CW, Noor AM, Marsh K. The changing limits and incidence of malaria in Africa: 1939-2009. Advances in Parasitology. 2012;**78**:169-262

[20] Müller MHB, Polder A, Brynildsrud OB, Karimi M, Lie E, Manyilizu WB, et al. Organochlorine pesticides (OCPs) and polychlorinated biphenyls (PCBs) in human breast milk and associated health risks to nursing infants in Northern Tanzania. Environmental Research. 2017;**154**:425-434

[21] Bouwman H, Sereda B, Meinhardt HM. Simultaneous presence of DDT and pyrethroid residues in human breast milk from a malaria endemic area in South Africa. Environmental Pollution. 2006;**144**(3):902-917

[22] Ntow WJ, Tagoe LM, Drechsel P, Kelderman P, Gijzen HJ, Nyarko E. Accumulation of persistent organochlorine contaminants in milk and serum of farmers from Ghana. Environmental Research. 2008;**106**(1):17-26

[23] Dalvie MA, Myers JE, Thompson ML, Robins TG, Dyer S, Riebow J, et al. The long-term effects of DDT exposure on semen, fertility, and sexual function of malaria vector-control workers in Limpopo Province, South Africa. Environmental Research. 2004;**96**(1):1-8

[24] Ennaceur S, Ridha D, Marcos R. Genotoxicity of the organochlorine pesticides 1,1-dichloro-2,2- bis(p-chlorophenyl)ethylene (DDE) and hexachlorobenzene (HCB) in cultured human lymphocytes. Chemosphere. 2008;**71**(7):1335-1339

[25] Chikuni O, Nhachi CF, Polder A, Bergan S, Nafstud I, Skaare JU. Effects of DDT on paracetamol half-life in highly exposed mothers in Zimbabwe. Toxicology Letters. 2002;**134**(1-3):147-153

[26] Turgut C, Atatanir L, Mazmanci B, Mazmanci MA, Henkelmann B, Schramm KW. The occurrence and environmental effect of persistent organic pollutants (POPs) in Taurus Mountains soils. Environmental Science and Pollution Research International. 2012;**19**(2):325-334

[27] https://www.jstor.org/stable/4186322 [Accessed: November 2, 2018]

[28] Agenda. Pesticides and poverty: A case study on trade and utilization of pesticides in Tanzania: Implication to stockpiling. Final Report 2006

[29] http://www.academia.edu/468465/Pesticide_pollution_remains...obsolete_pesticides_at_Vikuge_Tanzania, 2004 [Accessed: November 2, 2018]

[30] https://www.ajol.info/index.php/tjs/article/viewFile/100178/89440 [Accessed: November 2, 2018]

[31] Agrow. World agchem market steady. AGROW 497, 9 June, 2006